The COSMIC THEORY of THERMODYNAMICS

The Creation and the Secrets of Our Universe,
the Ultimate Being of Our Existence

Trafford
PUBLISHING®

Order this book online at www.trafford.com
or email orders@trafford.com

Most Trafford titles are also available at major online book retailers.

Illustrated and Edited,
Cover Design, Design and Photography
by Dimitrios Jim Vassiliou.

The information contained in this book is intended to be for educational purposes only.
The intent of the author is only to offer information of a general nature.
The author and the publisher are in no way liable for any misuse of the material.
The information in this book is the author's own point of view, and beliefs.
It is in no way intended that all-previous founding's are wrong.

Printed in Victoria, BC, Canada.

ISBN: 978-1-4251-8239-7 (soft)
ISBN: 978-1-4251-8240-3 (e-book)

*Our mission is to efficiently provide the world's finest, most comprehensive book publishing
service, enabling every author to experience success. To find out how to publish your book, your
way, and have it available worldwide, visit us online at www.trafford.com*

Trafford rev. 11/06/2009

 www.trafford.com

North America & international
toll-free: 1 888 232 4444 (USA & Canada)
phone: 250 383 6864 ♦ fax: 812 355 4082

I BELIEVE THAT IF THE WORLD CAME TO AN END, AND THERE WAS THE SLIGHTEST POSSIBILITY THAT THIS BOOK COULD SURVIVE THE DISASTER, AND NOT BE DESTROYED; THEN IT COULD BE USED AGAIN IN ANOTHER LIFETIME.

I HAVE DISCOVERED A NEW THEORY OF PHYSICS BEYOND STANDARD MODEL. I CANNOT EMPHASIZE ENOUGH THAT THIS NEW THEORY OF PHYSICS IS BEYOND THE STANDARD MODEL THAT SCIENTISTS ARE USED TO WORKING WITH TODAY.

I STARTED TO WRITE THIS BOOK IN 1955, AND DUE TO THE DIFFICULT NATURE OF THIS BOOK IT WAS COMPLETED IN THE YEAR OF 2002. I HAVE DEDICATED MY WHOLE LIFE TO THIS BOOK. I COMPLETED EXCLUSIVELY THE TOTAL WORK OF THIS BOOK.

I, DIMITRIOS JIM VASSILIOU AM THE AUTHOR AND THE EDITOR OF THIS BOOK. I AM A RESIDENT OF TORONTO, ONTARIO, CANADA.

CONTENTS

TABLE OF FIGURES

AUTHOR'S HISTORY

THERE IS ALMOST nothing that I did not do in order to succeed and make progress. I was insulted or I cried for any little thing that went wrong. I wanted everything to be perfect, all to be right, although I knew many a times that it could not be so. Since my years of childhood, I was interested to learn more about the planets, stars, and mathematics if possible. Live in peace and good health, away from trouble and the wrong ways of doing things. During the summer at nighttime, I was watching and admiring the **starlit skies** trying to come up with something to be related with our existence. I wanted to find out a theory in the science of physics, to explain how the universe started to exist. What was the **Ultimate Being of our existence?** This would concern all of mankind. At the same time I began to realize that I had limited economic ability. Thinking day and night, while sitting, walking or lying in bed, I was discovering there were more and more things to **look at.** How this world of nature had been founded and existing? Where was the huge power coming from, so that the stars and galaxies in the universe could last for billions of years? And how were they moving around? I was profoundly concerned of the Aristotelian philosophy and all his ideas about our existence.

I was pleased to study the fundamental issues being unsolved by the scientists', physicists and theorists. Those being mass, temperature, light, electric current and space, etc. I read different books and encyclopedias, and learned quite a few things of great interest for my perspectives. Among them was the meaning of the interesting words for me; for instance, temperature, thermodynamics, dualism, metaphysics, and others.

"The Consolidated Webster Comprehensive Encyclopedic Dictionary" explained the meaning of those words in detail consecutively found on pages 230, 455, 744, and 750. Very interesting explanations were provided for all of these words indeed. In 1955 I started to write **"The Theory of Thermodynamics"** and their relation to the **"Cold-dynamics".** Problems though of human concern, like marriage and growing children, had slowed down my efforts to continue and complete this theory. But a few years later I was back again to work. Since I was getting older I was able to do more and much better work to edit and finish my theory. I could find answers closer to the very important issues for my theory. I could also call it a cosmic theory of physics, which dealt with all the physical events and phenomena in the universe. I had studied the philosophy of Aristotle in my twenties and the philosophical ideas

1

and mathematics of Dr. Einstein in my thirties. This gave me more power to continue that important work. I wanted to hear later, that this theory was worthwhile, and be accepted and appreciated from those who might read this theory. Especially from responsible historians, and other scientists, physicists, and theorists. Let us not forget, we all probably know that humans differ in their mind. Not all of them accept everything that others are willing to accept. And this is natural. I will try to do my best in explaining how in my view I believe I have discovered most of the secrets in our universe, including the "Ultimate Being" of our existence. This in turn means the substance that has made the universe.

Always in the past, and still to this day, I like to read newspapers and different books which are interesting, including encyclopedias. Always learning something new. It is not a secret. It is true for anyone who wants to make his own structure of an important theory. He has to be aware of more scientific work done by physicists, theorists' and many others. It is true that my primary concern was the conception of the **"ULTIMATE BEING"** in our universe. I was thinking most of the time about the meaning of the temperature, which in my opinion it should be something that exists in our universe forever.

I knew that it starts from the absolute zero and goes up to millions of degrees; also that it was either cold from zero Celsius and down, or getting warmer and hot from zero Celsius and up. A temperature of two kinds, cold and hot and when it was too hot it would produce **ENERGY.** I will say much more about it in the next chapters. This cosmic theory of thermodynamics informs us that the absolute temperature includes both the dynamic properties of hot and cold in general. Thus, **"Thermodynamics"** could also be called **"Thermo/cold-dynamics"**. I have the same opinion just like other scientists, that the science of mathematics is very important for many big problems in this world including our universe and our existence. Up to now I believe that the equation to link all the forces existing in the universe have not been found. In the book **"The unexpected Einstein",** on page **224**, it talks about Einstein's goal of unifying gravity with other forces of the universe. I will try later to answer this very important issue as well as many others existing in this theory of cosmic physics. While I was writing my theory I found out and was aware that the rules of this nature were never changed. The nature was following the same route to reach its own destination. **"This cosmic theory of thermodynamics"** (cold-dynamics included) is repeated many times in my book; this happens because these words hold the stakes of the ultimate "being" in our cosmos indeed. Finally this temperature of hot and cold is the main cause that made my theory to be beyond standard model.

FOREWORD

THEORIES OF COURSE are discovered by human wisdom. I have started my cosmic theory of "Thermodynamics" in relation to cold-dynamics from a good point of view, according to my own perspective. I have directed this theory through a good path of comprehension for all those interested in reading this "COSMIC THEORY". I believe I have followed the right direction of this important cosmic theory that eventually may lead to changing other prior theories of physics in general. In my belief thermodynamics and the related cold-dynamics hold the main issues in exploring and explaining many cosmological events and phenomena in our universe. This in turn leads to the solution (and it is the most significant point of this theory) to many a problems relating to physics and metaphysics, including the puzzled problem of the general equation of the Grand Unified Theories; of the strong and weak nuclear forces as well as the gravitational and electromagnetic ones. From the very beginning of this cosmic theory, my conception is concentrated to a prevailing cold-dynamic power in our universe over a thermodynamic one in it. And it was the corner stone to discover many other problems in our cosmos.

THE FIRST PART OF THE COSMIC THEORY OF THERMODYNAMICS

CHAPTER 1

HOW I IMAGINE THE UNIVERSE

THE PHILOSOPHY OF Aristotle strengthened my will to follow good ideas about the universe. The universe in my opinion could be nothing else than what it is. An abstract image of an immaterial volume of space made up of light and dark representing the hot and cold temperature which expands and/or contracts the volume of the space. This hot and cold again is being transformed to masses of liquids, solid objects, gas, etc. And vice versa. To do this by necessity the universe acquires energy and this comes from high and low degrees of temperature. So again by necessity in primordial epochs a pre-existing immaterial mass of unimaginable solidity from darkness, hotness, and coldness brought about the big bang force of explosion; and generated the space of light of all posterior lights and spaces after. Local and distant thermal balances made out of planetary systems in a microcosmic and macrocosmic world of nature had been developed all over. These planetary systems and all motion in the universe generally are developed as a result of electromotive forces locally and distantly between the cold-dynamic and thermodynamic powers. An unequal amount of cold and hot power in the same absolute temperature makes one run after the other forever. They intend to bring about a balance in space, but of course in my opinion this could never happen. Later on in the book, I will explain all of it. Now if it were a materialized universe, how would it have been made?

Where would the source of power be coming from? I am imagining an immaterial universe of a volume of space, consisting of light and dark representing the hot and cold temperature. A finite universe inflated and deflated perpetually in the EONS OF TIME.

I started to write about how I imagine the universe in such a way, because I had made my mind up that the universe is comprised of galaxies with billions of stars radiating huge amount of energy every second in to the interstellar space using millions of degrees of temperature from the mass contained inside themselves. But then I thought that temperature is not always hot, it can also be cold temperature in space; so a relation could exist between both of them. I thought that wind forces are developed between hot and cold temperatures of two different regions in space. And how about the other forces in the universe? These are electro-magnetic gravitational strong and weak nuclear forces. How were they developed? Was that the result of hot and cold difference in temperature? Yes it was. I had no doubt that I was discovering fundamental issues of the universe. I began searching encyclopedias to find out the deep meaning of some interesting words like temperature, thermodynamics, duality, metaphysics, and many others.

In "The Consolidated Webster Comprehensive Encyclopedic Dictionary" on page **744,** it states that **temperature** is the state of a body or of a region of the earth with regard to heat, the degree or intensity of the heat effects of a body. On page **750**, it states that the word **thermodynamics** is that department of physics which deals with the conversion of heat into mechanical force of energy and vice versa. On page **230**, the word **dualism** states that it is a two fold division; a system founded on a double basis in belief of two fundamental existences; the belief in two antagonistic supernatural beings, one being good and the other evil; the philosophical exposition of the nature of things by the adoption of two dissimilar primitive principles not derived from each other; the doctrine of those who maintain the existence of spirit and matter as distinct substances, in opposition to idealism, which maintains we have no knowledge or assurance of the existence of anything but our own ideas or sensations. On page **455**, the word **metaphysics**, states that science which seeks to trace the branches of human knowledge to their first principle in the constitution of our nature, or to find what is the nature of the human mind and its relation to the external world; the science that seeks to know the ultimate grounds of **being** or what it is that really exists, embracing, both **psychology and ontology**.

There and after I had more assurance that the temperature with its potential source of energy was the essential, predominant, controlling factor and **the unique element** of our universe that we could not go beyond. The next chapter will mention more about it.

CHAPTER 2

ABSOLUTE TEMPERATURE AND THERMO/COLD-DYNAMICS

IN PLAIN WORDS of science, I may say that matter contains energy, and this energy comes from the temperature inside the matter. So mass is a source of energy and this is something that we should believe in. I am talking about absolute temperature, which in Celsius instrument the bottom line is **-273.16** degrees and **-459.67 in Fahrenheit**. This is as I have learned it at school. **Thus mass** is converted to energy and energy is the absolute temperature itself. I would say then that every piece of mass in the universe, which might be in solid, liquid or gaseous state, including stars and galaxies, should be nothing else than concentrated temperature. Now there are two kinds of temperature, they are hot and cold. I call them thermodynamic and cold-dynamic. The two of them are not exactly equal in power. The first has the less and it will act starting from zero Celsius and up to expand the universe, and the second with more power starting from zero Celsius and down to contract the universe. I call them thermodynamic and cold-dynamic or THERMO/COLD-DYNAMICS. More details are in later chapters. This difference in hot and cold power in the same temperature is the corner stone about everything concerning the whole universe; it is the sequence of events of how this universe is developed and working perfectly to reach its own destiny. The whole universe evolution being expanded and contracted is based on that unequal amount of cold and hot temperature. The energetic light and dark in space represents hot and cold temperature of two kinds in one entity. All of us know that temperature flows from a warmer object of mass to a cooler one, or it could be flowing between two different spaces in the universe, and one of them would have higher intensity of temperature than the other.

This is a clear meaning of how the thermo/cold-dynamics are related and working in the universe, making changes from one form to the other. It is of great importance to talk about this significant branch of the physic science. The object of mass in the last analysis will produce nothing else different than a dynamic coldness or hotness either to contract or expand the universe. The atoms inside an object of mass would never get exited if the temperature is between absolute zero and zero Celsius. From zero Celsius they start to get exited and in higher temperatures they will start to oscillate in a higher frequency and produce energy, which energy is a speed of light emission in square meters. This speed of light acts as a positive current.

Any object of mass close by with less temperature and negatively charged will be attracted. This is an important event and it will profoundly illustrate the mysterious secret of gravitational forces, which is the same force of attraction that has made the planetary systems of the atoms and the stars and galaxies. The positively charged object with the higher temperature is pulling around the negatively charged object of mass with the less temperature. This is working like a force of attraction. More details will be provided later. The distance and the density in space and some other factors will determine the power of the force of attraction. This is like the performance of an electric machine, but let me talk about this in detail in a later chapter. For now let us not be mistaken, this force of attraction between heat and cold could cause an attraction of heavier bodies of mass like the earth towards the sun. We are now again deeply concerned with these cosmological events of the highest importance. These events indeed reveal how our universe is working and releasing its huge energy for billions of years.

THE DEEP MEANING OF THE ABSOLUTE TEMPERATURE IN HOT AND COLD OR LIGHT AND DARK

I have the opinion and I believe that absolute temperature is the ultimate analysis of the essence of the births, endings and rebirths of the universe. It is all of our existence. It is a very essential constituent of our universe; it is also a predominant and controlling factor and the unique element of our universe that we cannot go beyond. It is withholding the huge, greatest, ever heavenly power. I strongly believe that we are living in an immaterial world. Theorists, philosophers, physicists and other scientists might think otherwise. What is the ultimate "BEING" of our existence? This means to me what is the substance and its absolute element, which ultimately created the universe? This substance is the temperature. This temperature is the sensation of how hot or how cold something is. I am confident about everything mentioned above.

Then all the activities of the temperature played and seen everyday should be immortal. Our whole universe is not made of matter; it is consisted of a plain absolute temperature of hot and cold appeared in different forms of nothingness. Existing and not existing! Seen and not seen! It is the light of life, and the dark of death, going back and forth. I call this temperature, thermo/cold-dynamic temperature and it seems to be our very "being" and our very existence. Evidently this is the central issue of the whole theory of thermodynamics. It is everything touched and not touched. The hot and cold of the pre-existing mass (see chapter 1) looks like an image of a marriage in the heavens, as it would be a mother and a father; the image of a father in the nature who came out as heat and light from a darkness and coldness, and the mother who was the same pre-existing coldness of mass, and she had the credit of being superior and more dynamic in power than the father. The more powerful cold and the less powerful heat when they were together, they were opposing each other, forcing the one to compress, while the other to expand. Indeed they represented a duality, made of two kinds of the same thing; that is a duality in one entity, or as it is described in the "Webster Comprehensive Encyclopedic Dictionary" on page 230, "A system founded on a double basis in belief of two fundamental existence, or even two antagonistic supernatural beings, the one good, the other evil."

When the big bang occurred simultaneously the leptons made their appearance in the skies and they had the image of the children who came out with an inherited thermo/cold-dynamic power intrinsically in themselves as that of their father and mother; as a result they produced a universe of the same ratio as that of the pre-existing mass made from an unequal in size hot and cold temperature forever. The same general forms states and changing conditions and phenomena existing and performed in cosmic times of primordial epochs billions of years ago, are now taking place in a similar sequence which resemble exactly like those previous events. The universe follows its path without changing it. This temperature also can be seen under the form of different colors of light, or even under the form of an object of mass, solid, liquid or gas, live and active or dead and inactive, and under any other condition continuously changing in the universe.

I believe that the duality state of temperature is responsible for many other concerned changes and performances, including the hot issue of how the forces were created in the universe, etc. Now then is the time to explain why I came to the conclusion that all the masses in our universe consist of a concentrated hot and cold temperature. In my opinion there would have never been a big bang unless there was existing a pre-existing mass consisted either of more cold and less heat or of more heat and less cold. I seriously and carefully thought and I found out that in our universe there is a motion; but how was this motion created? I had figured out that changed temperatures in different regions of the space in our universe make the winds; and that is because of the difference in temperature between the two regions in the space. This situation is created because of a higher or lower temperature in one of the two regions. So I thought that if the amount of the hot temperature was higher in the pre-existing mass the universe would be probably expanding forever and we might have an open universe. But if the amount of the cold temperature was higher then probably a finite universe could exist, being expanded and contracted. And if the amount of hot and cold temperature was the same, then the universe would not move at all; it would probably stay idle and in a permanent stability.

I think that in the last case there could not be either expansion or contraction. **The inference is that there always should be a coexisting motion in an open or closed**

universe. All run and vary continuously and perpetually in this universe in an un-equal state of expansion and contraction in different parts in the interstellar space. I think I found out exactly what was the amount of the cold-dynamic and thermo-dynamic powers in the pre-existing mass, and how much the universe should be ex-panded and contracted in order to be working perfectly and have a perpetual pulsat-ing universe in the eons of time. There are more details on these cosmological events that you will find in later chapters. For now again I would say that if it was possible to have a different ratio in the amount of the cold-dynamic and thermodynamic powers than that exactly which is existing right now in our universe, then I believe that the whole picture of our cosmology would have been radically changed.

I do not think that anyone would really believe the almost uncountable num-ber of applications that the **dual** of hot and cold has to offer in our misleading whole world of nature. In the light of these applications I believe I have discovered cosmo-logical issues leading not only to the ultimate being, but also to the science of meta-physics as well. These important cosmological issues might have challenged the scien-tists of all times. I believe now it is the time to talk about and answer these important issues for our humanity. I found out in the "Consolidated Webster Comprehensive Encyclopedic Dictionary" on page 455 that "METAPHYSICS is the science that seeks to know the ultimate grounds of **being** or what it is that really exist, embracing both **psychology and ontology**". I believe in my opinion I found the "DOCTRINE of ULTIMATE BEING", and this should be a part of the metaphysics.

Scientists now could investigate and explain the nature of our exis-tence for themselves. I must say only that in my opinion, in this misleading world of our nature, we think we are something but we are nothing, because we came from nothing, which we think is something! In our universe there are "THERMODYNAMIC TEMPERATURES" which are related to the "COLD-DYNAMIC TEMPERATURES". The first will bring forth heat, light, energy, life and existence, while the second will bring forth cold, dark, inert and death. It is an existence of the day in the universe, and the dark of the night. The life will start with the "Big Bang" and it will terminate when the mean temperature will reach the coldness of the absolute zero. This will be talked about later. The heat and cold are running one after the other taking us either to the light of live external world, or to the dark and Hades of a metaphysical world.

Cold and heat are intrinsic properties of the temperature as I did mention it before. The hot and cold in one entity temperature will never cease to exist in eternity. It will exist in any state and form in the infinity of time, of the primordial, medieval and posterior epochs. With our human wisdom we may be able to find out a serial number of secrets and cosmological events in our universe. For example how is it functioning perfectly in a magnificent way of getting through its path of destiny, returning back, restarting and so on? The temperature is transformed to billions of changing colors in the space, starting from dark blue at the beginning of the big bang creation (no light wavelength) to the very end of blackness and coldness of absolute zero (light wavelength infinite). This is always an unsteady temperature because of the extra amount of cold in it, and it will never let it balance in the space any time, anywhere in eternity.

A NEW THEORY OF PHYSICS BEYOND STANDARD MODEL

As the author and editor of this theory I would like to explain a few things so that the readers of this book have a better understanding. The first thing to mention is that this whole theory of mine, is mostly based on my imagination. How can I solve the difficult problems concerning the universe and find what I'm looking for? For these problems I have read many books and encyclopedias. I would have to say that the most interesting books for me were the "Aristotelian Philosophy", "Building the Universe" which was edited by Christine Sutton in 1984, "The Big Bang, The Creation and Evolution of the Universe" by Joseph Silk in 1979, "Einstein's Theory of Relativity", many encyclopedias, and the "Consolidated Webster Comprehensive Encyclopedic Dictionary". I never copied anything out of these or any other books. I only made reference to certain findings from other authors and editors as to what they did or said about something, in order to compare their work and explanations with my work on the same matter.

English is my second language, and it took me hundreds of hours to find out the meaning of too many difficult words for my new theory. I strongly believe that this theory, which is beyond standard model, is a revolutionary one. It is a theory of putting things together in a different way than many other similar ones. The purpose of this theory and the purpose of my perspectives are to show how this universe was created.

What made the stars and the galaxies? The book, "The Big Bang" by Joseph Silk says on page 201: In some old stars, a sudden outward eruption of neutrinos from the core can master enough clout to blast the star to pieces in a titanic cosmic explosion known as a supernova. My question is how does that happen? I think I know the answer and I can prove it later on in the book. The newspaper "The Toronto Star" on February 5,1996 said that technicians at Sudbury Neutrino Observatory are helping to build the world's largest and most sensitive detector of sub-atomic particles. Deep underground near Sudbury, scientists hunt for elusive neutrinos. Maybe they are possessed of mass, maybe just energy. Most certainly the smallest, most quizzical and mysterious of the fundamental building blocks that make up the atoms of the universe.

I thought about these problems day and night; these same problems that the physicists encountered from decades ago. Christine Sutton, editor of the book "Building the Universe", says on page 125: "What sets the electron's charge at 1.6 * 10^{-19} coulombs? Why is its mass 0.51 MeV and not 5 MeV? And also the negative charges "in" the electron all repel each other, so what then holds the electron together?" These all are repeated in more detail later. And again, using my new theory in physics beyond standard model, plus my imagination, I tried to find a solution for all these things. I thought about the four forces in the universe, which are mentioned on page 5 and page 177 from the book "Building the Universe". I asked myself what caused these forces to be created in the universe? Christine Sutton says, " Each force in its own way is quite fundamental to our existence. Moreover, it is the precise nature of the forces that makes the universe the place we observe. But particle physicists seek more than just an understanding of each force in its own terms. They hope one day to come to a single theory that encompasses all four forces, and thereby all the diversities of our universe. Such a goal may still be a long way off."

I was anxious to find out and see if all these problems could be solved once and for all. I knew I had a good reason to be hopeful because I was confident that my theory was the right one and worthwhile to answer almost anything unfinished in our classic physics. Is this goal mentioned above a long way off? You may see soon in this book that it might not be. I think that the particle physicists and other scientists are working hard to solve the problem of reducing the four forces into one, or find the structure of the atoms and probably answer many other problems such as how they should calculate the rest mass energy of each one of the leptons separately, or how they could also calculate the electric charge of the electrons orbiting their nuclei, and all others mentioned above. I think that because they are using our standard classic physics they might not reach satisfactory solutions very easily. This of course is my opinion. And going furthermore, it would probably be more difficult to answer about the gravity of Einstein, or about the existence of a materialized universe and so on. In my view, I think that I have enough evidence that there do not exist gravitational forces, or a materialized universe in this cosmos! The earth is not orbiting around the sun by gravitational force.

I strongly believe it is attracted because of a positively charged sun with a much higher power as a result of its huge mass and amount of a higher temperature; and a negatively charged earth with less power as a result of a much less amount of mass and a lower amount of temperature. Indeed, I also believe that exactly the same thing happens with the electrons orbiting their nuclei. You will see more details about these facts later. The other striking and very important hot cosmological issue is where and how the leptons can be found? What they are consisted of? How they are linked together to build quarks, protons, neutrons and more? These essential and infinitesimal particles that could build a whole universe are found in chapter

6, with all the details concerning them. The four forces in the universe are one of the most concerning problems in our universe. Christine Sutton as it is mentioned above, said that those forces are fundamental for our existence. In my part I have said before that cold-dynamic and thermodynamic powers of temperature are the cause that make everything to move in the universe. That is to say, the cause to make the electron orbit its nucleus is the difference in cold and hot power between those two (nucleus and electron). The electron is cooler and the nucleus warmer. The mass of the electron is infinitesimal compared to that of the nucleus and the distance in meters between themselves is extremely small, so that the power of the force of attraction imposed by the nucleus toward the electron is overwhelmingly strong, and it is called strong nuclear force.

Now in my view, I have reached the point to say that I am positively sure that this strong nuclear force is very much related to all other forces and does the same thing as the other forces do. In the external world this nuclear force is operating with huge masses and high temperatures and long distances, but it is exactly the same force that the sun is pulling the earth around it and the smaller in power galaxies are pulled around the more powerful ones. I will go into more detail about this in other chapters of the book. So, only one force in the universe is acting as gravity, electromagnetism, and strong and weak nuclear forces. Finally I think that only this new theory of physics beyond standard model could make the difference and solve the most difficult problems of our nature. I believe I have more evidence to show that this is true, and in my opinion I can prove it.

ABSOLUTE ZERO AND INFINITY

The science of mathematics helps to find out the equations leading to the solution of the Grand Unified Theories using the figures of absolute zero (0) and infinity (∞) that both seem that they are not actual numbers which could be used everyday. I would say that both of these are numbers and they are very important concerning the whole universe. If you multiply the absolute zero (0) by infinity (∞) it will produce a number (A) which (A) will symbolize any quantity or capacity of whatever immaterial substance including active or inactive objects of mass, volume of space, light and dark, or any abstract kind of number or whatever other form of numbers etc. Further performance will show that zero divided by zero (0/0) would be impossible (it can not be determined). For instance it could be 1.33 or 9999. Finally, an abstract number (A) or quantity of anything, divided by zero (A/0), will produce a quotient, which will mean infinity (∞).

That is: $0*(\infty)=A$, and 0/0=impossible (or it cannot be determined). $A/0=(\infty)$, and $A/(\infty)=0$. These equations overwhelmingly describe the magnificence of the static dynamism of the figure of the absolute zero (0). The phenomenon of the existence of our universe is a mystical problem to our mankind. It is described by abstract senses, like space of infinity (∞) or absolute zero (0), and by dynamic masses of heat and cold. We could say that it is a visible or an invisible universe without this sentence to constitute an apparent false statement. The absolute zero (0) gives birth to an apparent materialized universe, but in essence because the absolute zero is nothing it will give something from nothing that it is nothing more than a changing dynamic temperature. Remember temperature means heat and cold. Heat and cold always have to co-exist in our universe; they are never separated. Absolute zero (0) and infinity (∞) are numbers, which cannot be counted. We are existing in a finite universe of infinity (∞) which is not countable, and when the cosmic time comes with an ending of our universe which will also be the beginning of another one again, we all are going to cross the point of an absolute coldness solidity in the line of a circle which I will call it an absolute zero or a point from a metaphysical World of Hades to a live world of light. Both Hades of metaphysics or light of life are consisted of concentrated temperature of nothing but this is not nothing. This is a something of nothing. This absolute zero (0) includes in itself a whole universe!

The first and main principle of my cosmic theory is that everything in our universe has a beginning and an end. It starts from absolute zero (0), goes to infinity (∞) and then returns back to the absolute zero (0). There is as much capacity in absolute zero in the metaphysical World of Hades, as much as there is in the finite infinity (∞) of our present universe. The absolute zero (0) is an abyss of the metaphysics enclosed in a circle. It is the room, which holds our entire universe. Our universe cannot and must not be expanding indefinitely; it is a finite and a closed universe. If it were expanding indefinitely, where would the source of power come from? There should exist an everlasting self-source of energy, and more importantly no absolute zero (0) and abyss of Hades and metaphysics would exist either, in order to encircle all that space expanded and all the masses if and when the time would come to have them returned back; everything in the cosmos would seem to grow up one way, never returning to where it came from. This situation would go on forever, and it would drug and carry away the entire sea of active and inactive masses in our universe, including us.

Absolute zero (0) and infinity (∞) are equal in capacity; both of them are floating in a bottomless and boundless ocean. The (∞) represents an external phenomenon of live appearance and motion of an immaterial thermo/cold-dynamic world while the absolute zero (0) represents the Hades of an internal and metaphysical world of a crystallized coldness and an infinite density. It can contain as much as the external world of a finite infinity could. The universe starts from zero that goes to infinity, and then it comes back to zero.

$0 = A / \infty$, $\infty = A / 0$ and also $A = 0 * \infty$.

CHAPTER 3

THE EPOCH PRIOR TO THE BIG BANG AND AFTER

THE EPOCH PRIOR to the big bang (see page 315 of the book "Big Bang" The creation and evolution of the universe by Joseph Silk - October 1979.) full of important details starts at the point where there is less expansion than contraction in the universe, when the day ends and the night starts; when the energy of radiation in abundance ceases, and the thermonuclear reaction in the stars and galaxies subsides down. Eventually in the end all these inert bodies of mass (dead stars, etc.) of condensed thermo/cold-dynamic temperatures will be exposed to the prevailing frigid coldness of the surrounding space, and they of course will still be stepping down in a dying process, until being entirely eliminated to a lifeless mass. The volume of the universe is shrinking, not expanding anymore. It is being absorbed by the cold-dynamic power of the extreme cosmic low absolute temperature and transmuted into a pure cold space. A previous energetic mass now compressed into a core of the smallest ever imaginable volume of the whole universe. It will be compressed very hard and with such a powerful pressure, consuming an amount of energy as that as the energy that has been released before to create a whole universe with the billions of stars, galaxies, and constellations that has lasted probably many billions of years.

I believe that the big bang indeed has occurred and there could not have been any alternative procedure that would have made the genesis of a universe in another way. In my opinion the big bang was a brilliance of a mighty force, of titanic dimensions, which shook the heavens. With a genesis and creation being in process, the day was born from the night and the space of light and dark. It was the first primordial light creation. The young globe of the early universe at the beginning was a chaotic state of agitation; a flaming space in the highest form of a thermonuclear storm. I emphatically believe that the leptons created were the most mysterious structure of the universe, famous illusive immaterial particles as we may see that later. I was thinking many times that they were the particles and blocks that produced a whole universe of stars and galaxies etc. Before the big bang occurred and moments earlier the universe was in the form of a sphere of mass as I did mention above. This was the pre-existing motherhood of the nature, the rigid coldness and hotness of the cosmic thermo/cold-dynamic temperature. That lifeless mass of the inner world of abyss and metaphysics had the heat and cold power in a degree of unimaginable dimensions, and the much more stronger cold-dynamic power in all parts of its mass (as a result because of its prevailing cold power) it was in the process of becoming more and more pressed to a point that it could not go beyond. That was the case. On the other hand the thermodynamic powers could not stand that increasing pressure indefinitely. The hot part of the sphere had no other way but to react.

The highest opposition should have come from the center of the sphere where the adamantine solidity or density could not be expressed by human words. The heat power source then in the core of the sphere which sustained the huge pressure of the cold mass, wanted to have an exit. At the same time an electromotive force had to be developing progressively between the warming up center and the outer perimeters of the sphere being in an absolute zero degree of temperature. This electro-motive force (EMF) or voltage being developed between two points as a result of difference in absolute temperature degree, got higher as the heat in the core became more intensive. It is exactly similar to a situation where there is an electric machine that electric resistance (R) in ohms and a current (I) in amperes flowing through a conductor, an (EMF) is developed because of a potential difference between two points. These developing situations were responsible to make the spark get started deep in the center of that heavenly mother that would build a whole universe after. Being now red hot inside the core it was getting extremely strong in thermodynamic power and it started to oppose seriously the antagonistic supreme cold-dynamic power pushing each other into the opposite direction although they were fighting before in other cosmic times for billions of years to form a balance of power in the space (like two in one) and as they are now and always trying to do. The thermodynamic power was in the process of beating the cold-dynamic power.

The lifeless mass was lit and the hot power had the upper hand. Moments before the big bang occurred, both cold and heat power masses were in motion, each one against the other until the big bang changed the whole picture. The critical point for the big bang when it occurred was when no more pressure was tolerated among these two powerful giants, heat and cold. They touched the end of the line on the circle which is the beginning again. They wanted to burst out to generate an energetic space of light and life and time; to give birth to leptons, atoms, stars, and galaxies; to make them orbit around in the form of planetary systems in local and distant thermal balances. The strength of the explosion of an immeasurable and almighty heat power was a deity of light which sprang out with an image of radiant splendor similar to that of Jesus Christ; the radiant and brightest ever blue beams around his holy face; the image of fatherhood coexisting with the image of motherhood. Thermo/cold-dynamic masses of power. This panoramic splendor of a grand magnificence was the image of an unaltered deity in our cosmos. The incandescent and flaming center of the sphere instantly gave momentum to generating a young globe of universe taking lightning and accelerated dimensions. This procedure was measured in picoseconds and the young globe was baking and boiling.

Frightful forces of the heavenly released energy of a power rate of light emission were spreading in every direction. An almighty big bang in force accompanied with collisions and thunders. Possibly big chunks of live mass with thermonuclear reaction in them were shaken away; and probably later these masses could shape stars and galaxies. Most of the space of course was filled up with leptons, that with their interactions were producing later the atoms and the heavier masses. The space of energetic life in our universe had started. It was the image of resurrection of Jesus Christ or the deity of light of all other posterior lights in our universe. It was a regenerated universe, which would remain perpetually pulsating and recycling in eternity, and always an infinite time accompanying it. This rapid development of the young globe of an embryo universe brought other events in the surface, of cosmological dimensions to be discussed later.

THE CYCLING OF THE UNIVERSE

The universe is perpetually pulsating in the forms of either expanding or contracting, "living or dying" in days of life and nights of death. This is because of the duality state of heat and cold of the thermo/cold-dynamic temperature condensed in one entity. The days of life is the active part of the universe, so long as it is expanding because of the thermonuclear action after the big bang in the masses themselves. The nights of death come later when our universe becomes less active progressively eliminating to a stiff coldness of density. Actually I am relating here to the cycling of the universe, which is completed in two great stages. The first stage starts after the big bang, which generates a new universe. The first stage is the day of the universe of the brightest part, the brilliant part, the splendor of genesis, the deity of the fatherhood of light and space and life, the heat of the temperature radiation and expansion, the positive side, the male or the thermodynamic part of a misleading world of nature with stars and planetary systems, the speediest action of a major volume of space with the great forces of attraction developed in micro and macro distances.

The second stage is the night of death, which is the inversion of the things that have happened before. It is the dying of the light and life, the increasing of the darkness, the side of the coldness and inertness, the diminishing of activities, the contraction of the space, the eliminating speed of kinetic energy, the dense cold and darkness in the abyss of a metaphysical world. Figure No.1 will show the day and the night of the universe.

The Big Bang, the mother force of a titanic explosion flaming and boiling with the highest pressure and temperature. Chaos and agitation in the first picoseconds in that primordial epoch.

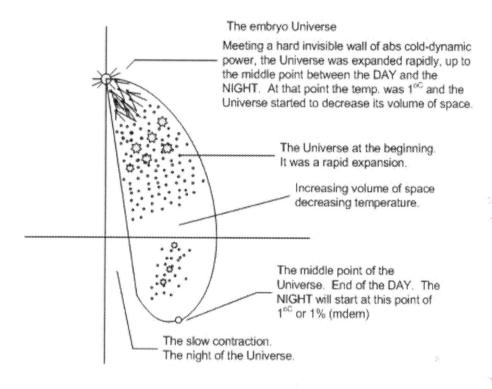

The embryo Universe

Meeting a hard invisible wall of abs cold-dynamic power, the Universe was expanded rapidly, up to the middle point between the DAY and the NIGHT. At that point the temp. was 1^{∞} and the Universe started to decrease its volume of space.

The Universe at the beginning. It was a rapid expansion.

Increasing volume of space decreasing temperature.

The middle point of the Universe. End of the DAY. The NIGHT will start at this point of 1^{∞} or 1% (mdem)

The slow contraction. The night of the Universe.

Figure No 1
Shows the Universe expansion and contraction diagram

THE STRONG AND TIRED ENERGETIC LIGHTS IN THE UNIVERSE AND ITS VARIETY IN COLORS

The pure and vivid light itself emerged at the same time when the big bang occurred. It was a very strong light, which I believe to be of a bluish color with a zero wavelength in it. It was made out from the part of the heat power of the pre-existing mass. That mass is consisted of so much heat and so much cold in percentages as I will explain it in a later chapter. The heat power is condensed intrinsically with the cold power in every piece of mass even in the smallest particles of leptons! In other words the hot portion of this mass is responsible for the generated pure light in the first picoseconds of the big bang. This light could not live alone in the space, it is penetrating deeply and rapidly, but almost instantaneously, it returns back to where it came from shaping the same particle or object of mass, which mass is made of heat and cold (see page 157 in the book "Building the Universe", edited by Christine Sutton in 1984). All the energy released in the whole universe is produced by the part of the hot percentage in the pre-existing mass. This light of lights from the big bang is never to reappear until a new generated universe is created again.

I believe that the masses at the beginning of the young universe had the highest density. These masses were in an active state of a thermonuclear reaction and later they shaped the atoms, the stars and the galaxies. The active atoms now in the stars and galaxies are the real and genuine descendants and/or secondary generators producing the energy and expanding the universe. All intensities or colors of lights should correspond to a shorter or longer wavelength. And also I believe that every color of light in the rainbow band of colors corresponds to an analogous intensity of a degree of temperature. The rainbow band should contain an infinity number of light colors that could not be visible and distinguished by the humans. In an absolute zero (0) degree of temperature the power rate of light emission is zero (0) and the color of mass is absolute dark. The wavelength becomes infinite. The opposite will be witnessed at the instance when the big bang will occur. Then the greatest power rate of light emission with no wavelength (pure light) will be generated. It will be rather a bluish color with a shining brilliance, probably never being produced in the billions of years of life in the universe. The most powerful light with the smaller wavelength travels a longer distance in radial and straighter lines than that which would possibly travel a weaker one. If both a strong and weak beam of lights are influenced by other objects of mass in the space while traveling through, the weak beam of light will incline and be deflected more easily towards the object of mass met in the space (see page 133 in the book "The Unexpected Einstein").

Ending up the whole volume of space is an energetic light of infinite number of colors, which are corresponding to a certain degree of wavelength, brightness or intensity of the absolute temperature.

THE TIRED LIGHT

An energetic light in an energetic space has a corresponding higher or lower power of going straight or has a curved direction in the interstellar space, depending on its shorter or longer wavelength that will be given by its source of power. The curve of the beam of light with a shorter wavelength will be less diverged from its course than that of another beam of light with a longer wavelength. See Figure No.2. In the long run it will follow a globular line of the universe, which I believe it has the shape of a sphere.

Fig No. 2

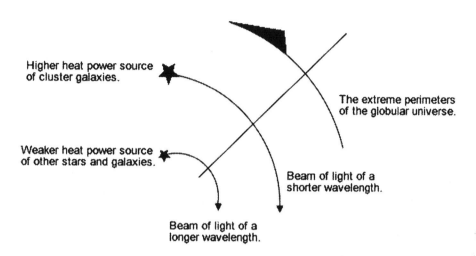

Higher heat power source of cluster galaxies.

The extreme perimeters of the globular universe.

Weaker heat power source of other stars and galaxies.

Beam of light of a shorter wavelength.

Beam of light of a longer wavelength.

There is always a chance where the energetic beam of light reaches a point in the interstellar space where it beams to cool off and contract. There at this point the thermodynamic power source that has produced this beam of energetic light has been defeated by the cold-dynamic power of the interstellar rigid coldness.

One could see that different energetic lights are created at different local and distant regions where powerful and/or weaker stars and galaxies, and other celestial bodies dissipate their energy of any power rate in to the vastness of the interstellar space. The light beams generated out of these sources, have a spiral direction towards the invisible wall of the extreme borders of the universe, while they are traveling around the globe of the universe. They will continue to do so in order to extend the volume of space, much as their power of energy permits them to do so. They will follow that direction, before they start to contract and fade away as a result of their weakness and lack of energy. They would probably not be able to overpower the resistance of density met deep inside the kingdom, of the cold-dynamic side of power. Like exactly what is happening to an electric machine with a current flowing through in a conductor of an extensive length and facing a stiff and increasing resistance in the wire. You may remember that the color of the space automatically is analogous to the temperature of this particular energetic light at that very moment. We should keep in mind that the speed of light of any source is always reduced through the density in any region of space. This is happening because density in the space is the mass itself that could be expanded and/or condensed.

The light beams or particles of mass, most of the time find a resistible space which could be a warmer, a cooler, a cloudy or any other kind of space of a higher or a lower density.

CHAPTER 4

THE TRINITY

WE NOW ENTER the part of the essence of the light, the dark and the space. When the big bang occurred and the light emerged out of the abyss of darkness, it was not only light, not only dark, but it was also space! All three were in one, like the image of "**the Mother, the Father, and the Holy Spirit**". All the above are a symbolical representation of the mystery of the trinity. The two in one (heat and cold) by birth together could never have existed without occupying a space at the same time. These two in one also could never be able to expand or contract or make any move without having the power of extending the space simultaneously. The image of a spiritual space, the divine spirit of propagation the spirit of God, the third person of the trinity animating with vigor to convey away secretly, as if by the agency of the spirit (see page 697-698 in the book "Consolidated Webster Comprehensive Encyclopedic Dictionary"). This image is the belief of communications by means of phenomena. Thus not only light, not only dark and not only space. All of them are one. All three must co-exist together, everywhere at any form or state, at any infinite time, which follow their moves in the eons of time.

The dualism of heat and cold pre-existed and co-existed together with a finite space between the absolute zero (0) and the infinity (∞). Looking deep in this cosmic theory of thermodynamics, mainly I am dealing with the creation of the universe. What is the ultimate being? And what is the mystery of the strong and weak nuclear force, as well as the gravitational and the electromagnetic ones? (See pages 5, 113-114, and 177 in the book "Building The Universe"). All these together govern the universe and it looks like it is living in harmony for many billions of years. I will try to look forward to find out more secrets of the universe, and by entering deeply into these matters, I believe I will be many times confronted and confused with mysterious and paradoxical phenomena. Studying our existing or pre-existing mass, which has an image of uncertainty and misleading world, I discovered the power of trinity "IN ONE". Now using my wisdom I am trying to find out more secrets and mysteries of this universe as much as I can.

As I have mentioned before the pre-existing mass of heat and cold in absolute zero (0), was taken as an image of a mother and father in a misleading world of nature; an image of deity of the dark and the light. That image of the father was similar to that of **Jesus Christ with the bright beams around his Holy face**. Out of these images a **necessity was arising for the creation of the leptons that created the Cosmos starting from the atoms, which are the families of the nature,** just like the children who grow up here on earth! These children were the essential particles, which had been linked together to build the universe, as I will refer to again after. They were not only small infinitesimal pieces of mass but they were also of "MALE, FEMALE AND NEUTRAL" origin, playing like crazy kids with one another until they grew up to atoms, which had created the families of mankind, and later had constructed the heavier bodies of mass and the final buildings of galaxies. **All of the leptons being made of mass,** which were nothing else than **a duality of heat and cold temperature in one entity**. So why is it that this **misleading** universe could not play the role of a mother and father giving birth to their children, and by growing progressively these children build up the whole mankind of the universe? Why do **cold and heat and space differ from the image of the Virgin Mary, Jesus Christ, and the Holy Spirit, who are living together with the mankind that they have created**? That is of course the total of masses of the planetary systems, stars and galaxies, earth and its human beings. **I say it is an image that is forced upon us as something that could move us deeply in our faith**! Using my human wisdom while I was writing this cosmic theory I seriously had this conception about the mystery of the trinity, and the symbolical representation of Virgin Mary, Jesus Christ, and the Holy Spirit. I had also directed my attention to the cosmological phenomenon of how could cold and heat together as a dualism be propagated without a medium of space communication? And more profoundly yet, how cold and heat firstly could even exist without both having owned a space to be self-positioned? That would be unfounded and groundless.

That space now takes the image of the Holy Spirit and that is why I am explaining all of it in this chapter. I believe in this almighty power of the "Thermo/cold-dynamic Trinity in ONE". The light, the dark and the space are moving and pulsating perpetually in eternity. Indeed, it is a unit of all three properties in one. The deity of this trinity governs the universe. It emanates from the absolute zero (0), which is the abyss of a metaphysical world. It goes through radiation and expansion to a finite infinity (∞) and it comes back to where it started. Running the circular line in a permanent pulsation never ending in the eons of time. Thus you cannot speak of light only or dark or space being alone, it is impossible you cannot even think about it, they are all in ONE everywhere at any cosmic time. This whole thing, like a HOLY TRIO is giving you an impression of a splendor and a panoramic mysterious image of a similarity of the human religion. The image of Virgin Mary is the "cold-dynamic power". The image of God is the "thermodynamic power", and the image of the Holy Spirit "is the space of the light and dark propagation". **ALL IS HERE AND IT WILL BE HERE AND WE ALL ARE PART OF THEM ALL.**

THE DEFINITION OF THE VOLUME OF SPACE

In my opinion "VOLUME OF SPACE" is a view created by expansive and contractive action of forces developed by heat and cold powers coming from a certain amount of a pre-existing mass in the universe, and which volume of space is starting from an absolute zero degree of temperature growing up into an infinite magnitude, and coming back to where it had started from. The mass, which contains intrinsically the heat power, will expand to a certain closed and finite volume of space while the mass again which contains intrinsically the cold power will contract the volume of space to where it has started from.

As a result, in this way an **infinity number of colors** in the space will be produced and one of these colors, taken, as an average will determine **what is the hour of either the day of light in the universe or the night of death**. This volume of space will have the ability to be expanded when its mean absolute temperature will be swinging from the zero (0) degrees of Celsius and up, and it will be contracted from the zero degrees of Celsius and lower. **The volume of the universe is a container of heat and cold**.

Naturally no extension can be obtained without the means to have it extended. Also that there is a "FINITE INFINITY" in the external VOLUME OF SPACE as much as there is in the eternal chamber of an absolute zero (0).

$$0=A/\infty, \; 0*\infty=A, \; A/0=\infty.$$

CHAPTER 5
ENERGY IN THE UNIVERSE

ACTIVE MASS

In my opinion active mass is the mass with a thermic condition of a higher degree than zero Celsius temperature, or from **(273.16)** absolute temperature and up. From that point and higher the thermodynamic powers should start to act positively expanding the universe. From the same point of temperature and down to the absolute zero degree the mass is acting negatively contracting the universe. The mass with a thermic condition lower than zero Celsius temperature makes the cold-dynamic powers acting as absorbing powers to take off the released energy by the thermodynamic powers.

These cold-dynamic powers are not expanding the space of the universe. The first is called **ACTIVE PLUS**, because of its ability to radiate and expand the space, producing even more space into the existing one. The second is called **ACTIVE MINUS**. What makes it active mass, actually is its atomic oscillation as the physicists have discovered, corresponding to a higher or lower degree of absolute temperature above the zero Celsius; and we understand that the higher the sub-atomic oscillation in an atom, and the higher the degree of its absolute temperature, and the higher its luminosity and activity to radiate and expand and vice versa. The active minus mass is called that because it is acting below zero Celsius without any motion.

THERMIC CONDITIONS

I would say that a thermic condition is usually a term referred to any object or particle of mass indicating its warmer or cooler state starting from the bottom low of the absolute zero degree and up.

THERMAL CONDITIONS

A thermal condition may be a term referred to a more sensible or distinguished condition of higher thermic ability of an object or particle of mass, starting from above the zero Celsius degree of temperature and up.

HEAT POWERS

A thermal condition in connection with any quantity of mass (usually with large quantities) and higher or very much higher temperature will constitute a heat power source like that of the stars and galaxies. I believe that the amount of heat power of any object of mass will be equal to the product of the total amount of mass of that object multiplied by the mean absolute temperature per lepton in itself. That is: HP = tM * mT; which means total mass multiplied by the mean temperature.

REST MASS ENERGY POTENTIAL

According to the particle physicists "Rest Mass Energy Potential" of any object or particle of mass will be the energy in electro-volts produced by this object or particle when it is totally annihilated.

I would think that this is a very important thing for the cosmic physics. Scientists succeeded in determining how much energy is released from an electron plus a positron. The electron being negatively charged and the positron positively charged. This has been obtained in the laboratories. The scientists used huge machinery to produce high electric energy; and by installing strong magnetic fields, they accelerated the speed of the flowing electrons and positrons coming from opposite directions in ducks, and made them collide head on. In this way the (e⁻) and (e⁺) were totally annihilated giving up all their rest mass energy that scientists succeeded to measure and find out that it was more than one million electro-volts produced by both of them, or about 0.51 MeV (millions of electro-volts) in each one of them two.

In a later chapter I will make it very clear how that energy is produced, and the very important thing that probably puzzled the scientists for decades, why this rest mass energy of both the (e⁻) and (e⁺) is over one million electro-volts and the rest mass energy of each one of them is 0.51 MeV? I think I have enough evidence to show why and how this thing is happening, and I am proud of it. You may find out that the rest mass energy is not the same to both leptons of (e⁻) and (e⁺).

Talking about very high energies produced by heavy machinery and consumed in order to annihilate a plain electron (e⁻) and a positron (e⁺), what magnitude of energy would be necessary in order to annihilate a piece of mass weighing a kilogram? You cannot even think of it. And if you want to lose your mind, try to remember back in the cosmic times when a big bang occurred and a whole universe was created, most of it in pure light and leptons in an interval of time probably measured in nanoseconds or picoseconds. How much energy was consumed after this happened? Note that the same total energy released for a complete expansion of the universe (from top high temperature to 1⁰ Celsius), will be consumed to contract the same universe from -1⁰ degree Celsius to -273.16. The total annihilation of a piece of mass is a phenomenon of producing pure light probably without a wavelength. Humans to a very small extent of course only accomplish this. That would involve (e⁻) and (e⁺), or even a proton (P) to be smashed. The scientists say that smashing an (e⁻) or a proton will produce gamma rays, which would be powerful and penetrating energy radiation. They go onto say that this produced energy from an annihilated electron will be used to reproduce the same electron back again. I believe that this performance is a breaking down and re-condensing heat and cold temperature. The leptons are infinitesimal masses of condensed temperature, as you will find out more concerning the leptons in chapter 6.

THE DENSITY IN THE SPACE RELATED TO THE VOLUME AND THE CYCLING OF THE UNIVERSE

Fig No. 3

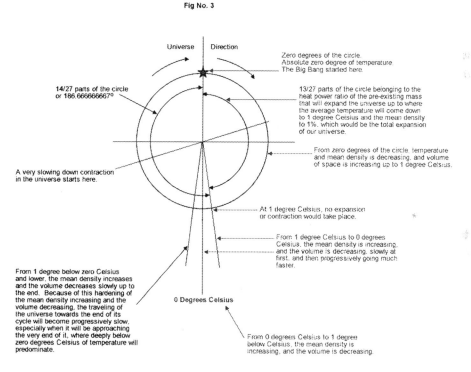

In my theory of thermodynamics I may use my own terms to express the density of the temperature, than that of other books in physics. Density means pressure or resistance within the mass itself. It is generated in the space, or in an object of mass like a gas, liquid, or solid. The density starts from 1% and goes up to infinity or to 100%. This is a characteristic factor taken into account with other factors, when the speed of light for the force of attraction is calculated between two or more objects of mass of different heat power in them. The maximum mean density (mDen) is accomplished when the absolute temperature comes down to the absolute zero degree. At that point the mean density reaches the 100% value or the infinity (∞).

Meanwhile the big bang would be the starting point where a decreasing mean density is beginning steadily and progressively to come down, until the average temperature in the space reaches 1^0 Celsius. At this temperature the maximum expansion of the space in the universe has been finished, and the mean density has reached the lowest point of 1%. The classic physics say that volume density and temperature are related to one another; and if multiplying (Vol) by (mDen) a constant number is produced. In higher mean densities the speed of the light should be decreased. Here are some characteristic numbers shown in Figure No. 3. From 1^0 Celsius to 0^0 Celsius while the (mDen) increases, the average temperature decreases. This is done because the leptons are still energetic up to the point of 0^0 Celsius. From 0^0 Celsius and down to the absolute zero degree, the whole space of the universe is steadily contracted; and like mentioned above the mean density is increasing while the temperature is decreasing. But this time this is not happening by the leptons, but by the cold-dynamic powers. That will slow down the speed of the contraction towards its end. Would the (mDen) ever touch down zero? In my view this could never happen. The lowest point of the average (mDen) as it has been said above, is the mark of 1%, and when this happens the average temperature in the space is 1^0 Celsius. In any other temperature the mean density is higher than 1%. The reason is that (you may wonder) there is no absolute emptiness in the space in a dead or live universe. There always exists an energetic light coming from a thermodynamic power source, or a dark light coming from an unmoved cold-dynamic power source. There is always dark or light in the space of the universe. This is a **NOTHING WHICH IS SOMETHING** or **IT IS SOMETHING WHICH IS NOTHING**. It is a ghost that needs a room or space to live in. The volume of space is a container of the cold and hot temperature. Refer to chapter 4, **THE TRINITY**. In Figure No. 3 you will find a circle diagram to show you the details of the mean densities, temperatures and volumes, at different cosmic times, while the universe is traveling to its own destiny. The 13/27 and 14/27 numbers of heat and cold ratios in the pre-existing mass are explained in chapters 6 and 8.

CHAPTER 6

THE LEPTONS – THE BLOCKS OF THE NATURE

THE LEPTONS WERE being born negatively, positively and neutrally charged. They grew up combining together in "TRIOS" and with the same procedure the trios made the quarks, protons and neutrons by using the magnitude of thermo/cold-dynamic temperature and the electromotive force of attraction during the earliest cosmic time when the big bang occurred. **The protons, the neutrons and the atoms were produced after**. There and after, the atoms will use their power of energy and will become local thermal equilibrium. Each one of these atoms consists of thousands of leptons and each one of these leptons contains a certain amount of condensed cold and hot temperature intrinsically in one piece. Thus the atoms themselves have a great deal amount of mass and energy when they are in the state of being active and exited as it was mentioned before in chapter 5 under the subtitle Active Mass. Also they could be ready to play a much more important role in general. They could become little planetary systems with the nuclei of the atoms doing the work of the stars, concentrating in themselves most of the mass, they are attracting the electrons which are being pulled and whirling around exactly as the planets are pulled by a force of attraction and orbiting around the star of the sun in our solar planetary system. (You will find details of this later.) The atoms look like families in their microcosmic domain and each one of them constitute a local thermal equilibrium. Their infinite numbers compose the whole universe. The immense significance of this event is the harmonizing and stabilizing performance of the whole universe locally and distantly.

The kinetic energy of the electrons on the other hand resisting the force of attraction by the nuclei makes them spin around with almost the speed of light for billions of years! For now I can only say that there could not be any stability in the primitive protium atoms without electrons circling around their nuclei. The reason is that the nucleus of the protium atom is the proton, and (remember) it has a neutral status by nature and I have named it the star of the universe. All other nuclei of any atom is a component of the proton. These nuclei could fly apart if they happened to be left alone by themselves. The same thing would follow if the sun in our solar system happened to be left alone in space. Thus like a necessity, the heavenly powers of "Trinity" of an adamantine solidity of the sphere of the pre-existing mass was broken and split into pieces, at that cosmic time of the big bang occurrence in order to create the misleading "World of Nature". The boiling and baking volcano of the embryo globe, as a result of the big bang event, in its first picoseconds generated clouds of vapor gases and/or showers of ashes of condensed cold and hot mixture. This heavenly thunderstorm was an unprecedented phenomenon of a demonstration of a titanic strength of force.

Sea of nebulae made out of incandescent gas of a bright magnificence, was the primordial image of the leptons being reacted with a successive complete annihilation and reproduction in that same space of a nuclear furnace! They were filling out the space as males, females and neutrals. These masses of leptons were in an intimate contact and top high temperatures and density. They were in their highest activity to constitute the foundation of heavier particles, stars and galaxies. It was not a materialized universe just made out of concentrated temperature of hot and cold. A universe of a deep vastness and interstellar space, full of stars, galaxies and other celestial bodies of mass countless in numbers. These leptons were the only basic elementary particles and the blocks of the nature from which all other particles and masses were made after. These particles were part of the doctrine for the origination and formation of the universe. Intrinsically they sustained the inheritance of the symbol of "Trinity". This Trinity was composed from hot, cold and occupied space. They appeared ready made coming out from the pre-existing mass. The appearance of these leptons was the start of the creation, and all this was done by necessity. So that the universe could proceed in changing forms and states, to follow its own destiny of being an illusive illustrated world of nature in harmony as it is seen and/or it is not seen today. Because of the immense strength of the big bang, the huge and total amount of the pre-existing mass of the lifeless sphere has been totally scattered and dispersed as a galactic nebulae, which filled up the embryo globe of space.

It was indeed a phenomenon of a thermonuclear furnace. This rare performance is not likely to re-occur for a long time yet to come. The total mass of the leptons themselves was the same pre-existing mass. The pure light of lights has been shown up, thus the image of God has been emerged out of the abyss of a pre-existing mass that it had been in a form of darkness and coldness and solidity. These leptons were and they still are the smallest pieces of an infinitesimal amount of mass. Within them there is a mighty of both thermodynamic and cold-dynamic power condensed in one piece. The paradoxical phenomenon is that these leptons with that imperceptible and infinitesimal mass had to be remained in that same form for billions of years in the universe. And they still are today. I profoundly believe that the pre-existing mass was entirely consisted of these spherical leptons as it is the whole live universe today itself. These things that I refer to as leptons are the very well known personalities of the electrons (e⁻), the positrons (e⁺) and the neutrinos (v). The electrons are being negatively charged and the positrons positively charged, both with a fractional charge of 1/3, the neutrinos are neutrally charged 50% negatively and 50% positively. All of them remarkably have a symmetry and precision of a delicate construction, as if they were created by a divine providence.

I think no human up to now has ever had the bright spirit to expect that these leptons could be shaped and constructed as I have discovered and found them myself and it is being explained in this cosmic theory of the physics. Theorists, particle physicists, philosophers, and other scientists in general, were puzzled and stumbled trying to solve these problems for hundreds and even thousands of years in the past. These famous particles were like descendants of the pre-existing mass. They were made as it has been said above from three kinds of gender: male, female and neutral. But also, each one of them precisely divided to three equal parts. Imagine now how much deeply in an unseen and in an inaccessible and impenetrable and indeterminable world, human wisdom succeeded to find out the end of the line. Leading to the mysteries and the secrets of how this inscrutable universe had started its creation, living and pulsating for so many billions of years, and not yet being close to an end. Figure No. 4 shows the electron (e⁻), positron (e⁺) and neutrino (v) construction. The three equal parts of thirds in the leptons are being condensed in the shape of a tiny sphere (a similarity of the globular universe), having an energetic hot and cold power intrinsically in themselves and all three of them I repeat condensed into one unit of an infinitesimal piece of mass being the lepton. Each and every one of the leptons will constitute a unit of electron or positron or neutrino.

Fig No. 4

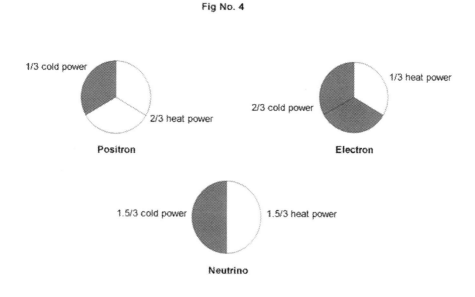

The amount of the power or the rest mass energy in electro-volts in each one of the electrons, positrons and neutrinos though is not the same as you may think. It is different all together. This is so because of the difference in the amount of cold and hot temperature contained in them. See Figure No. 4. These leptons (the blocks of the nature) as I said before appeared at the same time when the big bang occurred and the absolute temperature was changing from the absolute zero to the top highest temperature, which according to the scientific estimation it should be about 5,000,000,000 billion degrees of Celsius per lepton, corresponding approximately to 5,000,000 electro-volts. What could this mean? It meant that the leptons where made of temperature, and indeed that were the case. These blocks of the universe were ready made by a condensed temperature of two properties of the lowest cold and the highest hot temperature at that primordial time at the beginning of the creation. Starting with the electron (e⁻) it was made of a condensed temperature with two thirds filled up with cold-dynamic power of the bottom low and one third filled up with thermodynamic power at the top high of the absolute temperature and both condensed in one unit negatively charged and having a fractional charge of 1/3. The same procedure is followed for the positron (e⁺) being created again at the same cosmic time and it was consisted of 2/3 thermodynamic and 1/3 cold-dynamic powers positively charged with a fractional charge of 1/3. It was the anti-particle of the (e⁻). Finally the neutrino (v) had been made by 1.5/3 cold-dynamic power and 1.5/3 thermodynamic power in the lowest and highest degree of temperature as well. All these leptons I repeat were coming from the pre-existing mass of cold and hot properties, and all of them following the path of the mystery of the trinity. A cold motherhood, a hot fatherhood and a space occupied by the leptons. But this is not the end of it. The leptons of three kinds were supposed to be linked together in trios to make heavier particles and have a fractional charge of 1/3 each. Everywhere the number of three was present. To produce one quark or one proton or one neutron, I would need 1/3 (e⁻ and e⁺) and 2/3 (v). That would be the right way to produce each one of them separately, but these ratios are showing a cosmic equation, which is indicating the proportion of leptons existing in the universe, or the total mass of our universe. So that this equation will be the guide to show how you may connect all the leptons existing in the universe to produce all the heavier particles of trios, quarks, protons and neutrons which would make the whole universe. You may see after, how the leptons were linked up together and make heavier particles and masses to shape the universe as it is seen today with stars, galaxies and other celestial bodies. They used of course their thermo/cold-dynamic power to make them attract each other during the first era of an embryo universe, and in a furnace of billions of degrees in temperature.

COMBINED LEPTONS IN A SUPREME ANALOGY TO PRODUCE HEAVIER PARTICLES WITH FRACTIONAL CHARGES. QUARKS, PROTONS AND NEUTRONS

THE COSMIC EQUATION ONE (TWO IN ONE)

It would be a mysterious secrecy of the nature for any human to learn how and why two neutrinos were linked up with one electron or one positron to build a heavier particle of trio either being negatively or positively charged as I am going to explain this after. If you look at Figure No. 5 you will notice that each one neutrino has 1.5 thirds of cold power and 1.5 thirds hot power. The electron has 2 thirds of cold power and 1 third hot power, and the positron has 2 thirds of hot power and 1 third of cold power, that makes all together 5 fractional thirds of cold-dynamic and 4 of thermodynamic power easily being attracted and that makes the heavier particle of a negatively charged trio. If 2 neutrinos and 1 positron were attracted they would make a positively charged trio. Each one of these trios has a fractional charge of 1 third. Those 2 trios now could be attracted to build a heavier particle of 6 leptons neutrally charged, and this of course could be easily attracted by another trio, either positively charged to create a positively charged 9 lepton heavier particle or negatively charged to build a negatively charged particle of 9 leptons.

You may now see odd numbers of leptons connected creating positively or negatively charged particles and even number of leptons creating neutrally charged particles and so on. The two neutrinos in each trio are used to grab an (e⁻) or a (e⁺) so that they become the guards and protectors of them. The 1/3 of (e⁻ and e⁺) and 2/3 of (v) in the universe makes it work this way perfectly; ending up in the construction of the quarks, protons, and neutrons without having a lepton more or a lepton less remaining at all in the universe by itself. All these things of course were made by necessity in this multiform world of nature. The curious fractional charges of the quarks will be cleared now, as well as the proton charge and positron charge. Christine Sutton, editor of "Building the Universe" says on page 124, "The study of the leptons themselves has led to more questions than answers, and we have learned more about the basic nature of matter by studying much more complicated particles such as protons, or by studying the interactions of leptons with these more complicated particles". On page 125 she continues to say " It is indeed a simple particle (the electron), yet it leaves us with questions to which we have no answers. What sets the electron's charge at $1.6*10^{-19}$ coulombs? Why is its mass 0.51 MeV and not 5 MeV? And there is an old question first pondered by Hendrik Antoon Lorentz in the early 1900s, which remains unanswered. The negative charges "in" the electron all repel each other, so what then holds the electron together? " Further down on the same page she continues to say "But the muon's mass is about 207 times the mass of an electron. Thus the existence of the muon raises yet another question. How can two particles alike in so many ways, differ so greatly in mass?" And on page 245 she says "The equality of proton charge and positron charge is forced upon us, as well as the curious fractional charges of the quarks and the zero charge of the neutrino." Christine Sutton also says on page 129, "How can we calculate from basic principles the magnitude of the electric charge ($1.6*10^{-19}$ coulombs) possessed by all the charged leptons? This question like the question of how to calculate the speed of light from basic principles is so intractable that it must be left for future generations of hopefully brighter physicists."

I am proud and I believe I have found the answers to almost all of these very serious problems of physics that had to be resolved by our particle physicists. The solution of all of these problems is now beginning to surface. If indeed there weren't any curious fractional charges of 1/3 between the particle and/or the quarks and nucleons, what circumstances could have arisen in our universe? Did we ever imagine what cosmological events could be developed in our universe? I believe that higher fractional differences could develop a war of catastrophic collisions among the particles and this of course would have precipitated the end of our universe. Now before going any further to the construction of the quarks, I will write a few words about the force of attraction. In short, force of attraction is developed by electro-motive forces generated between cold and hot masses. This procedure is being done in a lightning speed locally and distantly. In extremely short distances and temperatures above zero degrees Celsius this force of attraction is extremely high. While in longer distances cooler bodies of mass like the planets could be attracted by the forces of attraction from other celestial bodies, heavy in mass and hot power like the stars. More details will be mentioned later in the book.

And now about the quarks, they are made of leptons of course. The more we research on them and the more we admire a multifarious and inscrutable universe. To think how these infinitesimal particles were created and how they are performing their job to build a universe is something that all of us should be admiring. It is like an intellectual performance by a divine providence. As I have mentioned it before on page 34, an electron or a positron could be combined with two neutrinos. So, in that case 50% of the neutrinos could be combined with the electrons and produce the negatively charged particles of a 1/3 fractional charge and 50% of the neutrinos to combine with the positrons and produce the positively charged particles of a 1/3 again fractional charge.

The conception of this analogy made me set the formula for a cosmic equation one (two in one). This conception of course has not been made at random. Each time I was thinking and trying to find the right solution of the proportions of all kinds of leptons to the whole universe. Page 694 in the "Encyclopedia Americana" 1997 edition says, "The mass of a proton is $1.6726*10^{-27}$ which is slightly less than the mass of a neutron, which is $1.6749*10^{-27}$ and 1836.15 times greater than the mass of an electron $(9.109*10^{-31}$ kg)". That showed me that the proton itself contains 1836 leptons; and that was all I wanted. Later I calculated the number of the leptons for each one of the three quarks of the proton and each one for the three quarks of the neutron. That was it. After a long and painful time, I inferred the entire space of the universe contains 1/3 of (e$^-$ and e$^+$) and 2/3 of (v). This was my beginning to start with the quark construction. I knew now the number of leptons to produce a proton should be 1836 and the number of leptons to produce a negatively charged neutron should have an extra negatively charged trio of two neutrinos and one electron, and the total leptons to make one neutron then should be 1836 + 3= 1839 leptons. But not necessarily all three quarks of the proton or of the neutron, should have the same number of leptons to make the quarks as you may see it after this. This was a very important method that changed in order to find the essence and the secret hidden behind the construction of these enigmatic quarks. With a different number of the particles of trios I could produce quarks being neutrally charged or up and down quarks. But I should be aware that all three quarks together should count either 1836 for the proton or 1839 for the neutron. This composition was a new method to reach the end of the line and solve this important problem for physics. This way the quarks were produced and grouped together in three as well as their fractional charges in thirds. The cosmic equation 2/3 of neutrinos (v) and 1/3 of electrons and positrons (e$^-$ and e$^+$) together in the whole universe constitute the right proportion. It was working perfectly. These leptons were binding together at the beginning of creation. Figure No. 5 illustrates how a negatively or positively charged particle of trio could be produced.

Fig No. 5

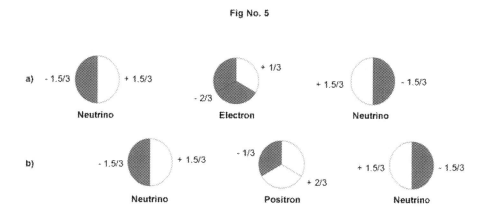

The positron-electron attraction is tremendously high in strength of force and these infinitesimal particles could possibly wipe out and annihilate each other if they came to a head on collision coming from opposite directions. That could have happened in the beginning, after the big bang, in very hot stars or even in a lab with accelerated speeds of electrons and positrons in opposite directions.

BUILDING THE QUARKS

I knew that particles as different as quarks and leptons should be grouped together in three, and their electrical charges should also be grouped together in thirds. It was a very difficult thing to build quarks in theory. I wanted to arrange leptons, so that all of the quarks in groups of three meet in general the terms and conditions in physics. Their fractional charges between the three quarks ought to be the same as those between the three different leptons in groups of three again! At the same time the total number of the leptons contained in each group of the three quarks for the proton, as well as for the neutron, be exactly the same, as they now exist in their physical condition. That being 1836 for the proton (P) and 1839 for the neutron (N). A very difficult situation was encountered indeed. I was busy trying to discover the solution to find numbers that could solve the problem. Each quark of the neutron should contain 613 leptons in order that the three quarks total 1839, and each one quark of the proton should have 612 to get the total of 1836 needed.

Arranging the leptons to build three quarks for the (P) and three quarks for the (N) was not easy. There was not a least common divider to fit all of them leptons in order to build the 6 quarks. I tried to find some numbers with the multiple of the particles of trios like the 3*204=612 and 3*612=1836, also 3*613=1839 or 17*6=102, but still I could not make any progress. No up and down quarks could be made and there was no way to build a neutron with 1839 leptons, or get close to it. But finally human wisdom made the miracle. I did mention in this chapter before that it was not necessary to have the same number of the particles of trios in each one of the quarks. Indeed the arrangement of the leptons to produce the quarks was hidden in the ratios 2/3 for the neutrinos and 1/3 for the electrons and positrons together. The mystery of the nature to produce the quarks for both proton and neutron was hidden right there. The cosmic equation of (two in one) could mean more for the physics because it produces not only protons and neutrons but also the quarks themselves, which make the protons and neutrons, and in fact the whole universe. It is the most important factor to build a universe.

This equation is: $1 = 1/3*(e^- + e^+) + 2/3 (v)$. And what (1) is this? That is: Either 1 quark, 1 proton, 1 neutron or 1 universe! In this analogy each quark of 612 leptons has 408 (v), 102 (e$^-$) and 102 (e$^+$), and in a proton of 1836 leptons the three quarks will contain 306 (e$^-$), 306 (e$^+$) and 1224 (v). That is 1836 leptons all together. You will see more details later. I believe these are all true and now it will be explained and you may realize it and agree why eruptions of super-nova stars are blowing out clouds of neutrinos as the astronomers have observed and called them neutrino stars. Einstein said, "God doesn't play dice!" But to me it seems that he does! The synthesis of the quarks construction has been found at the same time when I found the cosmic equation (two in one). The unstable neutron (N) on the other hand will consist of 1226 neutrinos (ν), 306 positrons (e^+) and 307 electrons (e^-). It is 1 proton plus one more negatively charged particle of "trio" of ($1e^- + 2v$) altogether 1839 leptons. It is a symbolic representation of trinity to watch again that there are groups of three quarks combined together with a fractional charge of 1/3 between them. But strangely enough the 3 quarks of a proton will produce a zero electrical charge for the nucleon itself. This is done in a mathematical and paradoxical way! It looks as something which is inconceivable but it is true and at the same time it should be postulated that these leptons of 3 kinds by their interactions at some primordial cosmic time were combined in trios and they have constructed the quarks in groups of 3 with a fractional charge of 1/3 among themselves. The progressively growing procedure of the sub-atomic particles of trio in order to produce a quark was in the form of: (-), (0), (+), or (+), (0), (-) of electrical charge, as it has been mentioned before. Out of the three quarks for the proton one was made neutrally charged, and that was the magic quark. It was the only one quark that all other up and down quarks were produced by this magic one. This could be done by adding or subtracting to this a negatively or positively charged particle of trio for making the other two quarks in order to produce the famous proton, the star of the universe. This proton will remain the central issue in producing all other nuclei of the atoms in the whole universe. This is done again by combining neutrons into the protons in extremely high pressures and temperatures. Also this nucleon of proton cannot live alone by itself, it has to be joined by a circling electron, or else it might fly apart like the neutrinos. The same thing would happen if the star of the sun in our solar system was isolated and being alone without its planets circling around it. Beautiful mathematical combinations starting from leptons and going on to construct heavier masses. All bound together in such a way of building our world of a misleading nature! Newton (1642-1727) once said: "Universe is having a logical structure which would probably be interpreted by the application of mathematical methods!" I believe this is true.

Future scientists and theorists could probably be able to find approaches of cosmic times and epochs of significant events that have been taking place in the universe by the use of mathematical equations, which are the laws of the physics. To discover and find out the combination of the leptons in groups of 3 and the production of the quarks in groups of 3 both with the same fractional charge of 1/3 between the leptons, and between the quarks was a remarkable and cosmological event in our universe evolution. It is important also for our mankind, as it is changing the course of physics to a new and entirely different path. I am very proud of myself to have all these achievements completed in this theory. Going to the construction of the nucleon of neutron now, this too is consisted of 3 quarks. The two of them down and one up as the particle physicists say. This is a negatively charged nucleon and it cannot attract and have circling electrons around it like the neutral nucleon of the proton. It is also intrinsically unstable and a little heavier than that of the proton, and as long as it lives in that state it will constitute a negatively charged particle, that could be linked with a proton or with a proton-neutron nucleon in very high temperatures and pressures. This could be accomplished because it is more negatively charged than the proton or proton-neutron nucleon. See previous chapters. I will explain now in more detail how the quarks started to be constructed at the beginning of the creation. An instant after the big bang where the annihilation and reproduction of leptons had almost calmed down, the combining of leptons in groups of 3 had started. This event took place while those crazy kids of leptons were running up and down in the space of the young universe like human kids today on earth before they grow up to make their own families.

They were fond of going together and they were helped by their (EMFs) developed, so they started to grow faster by making new heavier and stronger particles. They wanted to make combinations of 3, as there were three kinds of them (e⁻, e⁺,v). This symbolized the deity of "TRINITY". They produced an immeasurable number of these sub-atomic trios, all being smaller than the particles of the quarks, and all of them of course had different signs of charges, being of (-), (0), or (+) like those sub-atomic particles of muon, baryon, tau, meson and others, which embarrassed our particle physicists when they discovered them in the labs. It was a rapid combination being done by necessity, a process of producing the sub-atomic particles. The prevailing top high energetic absolute temperature, and the huge density made this procedure much easier to be made. The ending of this immense work came to an end at an instant when no more leptons were left in space alone.

The reason why none of them remained alone is because the number of all existing neutrinos in the space and the number of all the existing electrons and positrons together in the space were 1/3 (e⁻ + e⁺) + 2/3 (v). That is 50% of (v) linked with all the existing electrons in the space, and 50% of the remaining (v) were linked up with all the existing positrons in the space, and as soon as there was not even one lepton left alone in the space all the heavier particles of trios would have been finished. Now this whole number of particles of trios would be used to make exactly all the quarks needed to link together in three and produce all protons and neutrons, plus all other combinations of the nuclei in this universe. A universe, which is made of a cold and hot temperature! How these enigmatic quarks now were constructed it is another thing explained right here. I have mentioned before that there should be 612 leptons in each one of the three quarks to make a proton of 1836 leptons, and 613 leptons in each of the three quarks to make a neutron of 1839 leptons. But this could not work, and have these particles of trios being connected and make up and down quarks with fractional charges of 1/3 so that later they could be attracted to one another and make up protons and neutrons. This way they would be all the same, neutrals for the protons and down quarks for the neutrons. It would be impossible to do the job without up and down quarks and with a fractional charge of 1/3 in all of them except the neutral or magic quark that could be attracted by either an up or a down quark. At first I was puzzled. But later I was very happy to find out that a different number of the particles of trio in each one quark of the proton or of the neutron would bring unexpected and cosmological results, and solve a problem which I think is worthwhile and the right thing that had to be so.

There was no other way. I insisted to find a way and solve this problem. I was a winner again. Now I had to figure out first how the neutral quark (the magic quark) had to be constructed? Keeping in mind the cosmic equation one (two in one), I knew each time there were attracted 2 (v) with one electron or 2 (v) with one positron to make particles of trios. These particles were neutrally negatively or positively charged. Also I knew that even number of trios would make neutral heavier particles and odd number of trios would make either negatively charged or positively charged particles of trios, depending on the trio to be linked with the particle which had the even number of trios. If that trio to be connected was negatively charged like this: (e- + 2v) the even number particles of trios would become negatively charged and vice versa. If the connected particle of trio was positively charged like this: (e+ + 2v) then the even number particle would become positively charged. Also I knew that all the three quarks for the proton or for the neutron when they were ready, they should be all together 1836 leptons for the proton, and 1839 leptons for the neutron. These things are all known and mentioned in previous pages.

The next thing that was remaining was to start combining the particles of trios starting from a positively or negatively charged trio (does not make a difference in what order they start) to produce the neutral or magic quark, following the row of (-, +) or

(+, -). At the same time I was counting the number of the leptons and the trios I was using. I was working with these numbers all the time and finally I did find out that 102 electrons should be connected with 204 neutrinos, and 102 positrons connected with 204 neutrinos again to make 612 leptons total for the neutral quark. Exactly the number I was looking for! And there you have it; I made a complete neutral quark, also known as the magic quark! There were all together now 408 neutrinos grabbing 204 electrons and positrons together to make a total of 612 leptons. I was very happy indeed. This magic quark was neutral because it had been finished with an even number of 204 trios sharp! And that was the only one quark to be made as neutral. No other quark either in the proton or in the neutron would be finished with an even number of trios. All of the other 5 quarks for the two nucleons, which are protons and neutrons, were made with odd number of trios. Indeed to make the second quark for the proton I had to connect 103 electrons with 206 neutrinos and 102 positrons with 204 neutrinos, a total of 615 leptons or 205 trios altogether. An odd number of trios and I made a negatively charged quark with a fractional charge of 1/3 and that would be easily connected with the neutral quark. The two quarks now negatively charged with a fractional charge of 1/3 were ready to be connected with another positively charged quark to make the proton. The star of the universe! For this third quark I took 101 electrons being connected with 202 neutrinos and 102 positrons with 204 neutrinos and I made an up quark positively charged with a fractional charge of 1/3 and a total of 609 leptons or 103 trios now. (Odd number of trios again but a positively charged up quark.) I was very careful that all the leptons I used to make the three quarks should be 1836, and indeed if you count all the leptons I have used for the three quarks (neutral, down and up) you will find: 612+615+609=1836 leptons.

You may remark here that when you have the magic neutral quark ready, you can easily produce other up or down quarks by adding or subtracting positively or negatively charged trios. If you add a negatively charged trio you have ready a negatively charged quark with a fractional charge of 1/3. If you subtract a negatively charged trio you have ready a positively charged quark, and so on. You may also remark that positively charged quarks have an extra lepton of positron and negatively charged quarks have an extra lepton of an electron.

For the neutron now, it is made of two down and one up quark but the total of leptons here should be 1839 leptons or 613 trios. And for me it was easier this time because I knew I could use any number of leptons to produce an up or down quark. So, what did I do? I made one negatively charged quark like the one I made for the proton. I took 103 electrons, which were grabbed by 206 neutrinos and 102 positrons, which were grabbed by 204 neutrinos, and there you have it; one extra electron in this quark and I had ready a negatively charged quark (down quark) with a fractional charge of 1/3 and 615 leptons altogether. Exactly like I did for the proton. Or I also could add a negatively charged trio (1e⁻ + 2v) to the neutral or magic quark and make the same thing. To this quark now I had to add a positively charged quark. But I was careful one of the quarks should have three extra leptons more than that of the proton in order to produce a negatively charged neutron with 1839 leptons altogether. So what I did was to make another quark positively charged this time but with 615 leptons again like the previous negatively charged quark with the same number of leptons. It was very easy! I got 103 positrons combined with 206 neutrinos, and 102 electrons with 204 neutrinos; just one more positron extra and I made a quark with 615 leptons, but positively charged now with a fractional charge of 1/3. There were two quarks now linked together neutrally charged. And finally I made the last quark that should be negatively charged to produce the negatively charged nucleon of neutron. For this I got 102 electrons linked with 204 neutrinos and 101 positrons with 202 neutrinos. 609 leptons in total, and that was what I really needed to finish the last quark, to have ready the negatively charged neutron with 1839 leptons in total. Again with one only extra electron to make an unstable nucleon and with a fractional charge of 1/3. See below a synopsis for the quarks.

Fig No. 6

A neutral quark

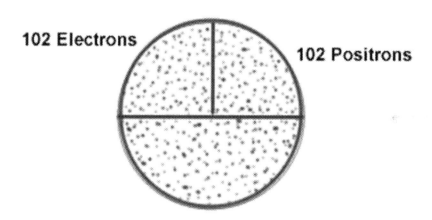

102 Electrons   **102 Positrons**

408 Neutrinos

A SYNOPSIS FOR THE QUARKS:

PROTON:	103e⁻	+	102e⁺	+	410	= 615 leptons of 1 down quark
	102e⁻	+	102e⁺	+	408	= 612 leptons of 1 neutral quark
	101e⁻	+	102e⁺	+	406	= 609 leptons of 1 up quark
	-------		-------		-----	= ----- ---
	306e⁻	+	306e⁺	+	1224	= 1836 leptons of 3 quarks
NEUTRON:	102e⁻	+	103e⁺	+	410	= 615 leptons of 1 up quark
	103e⁻	+	102e⁺	+	410	= 615 leptons of 1 down quark
	102e⁻	+	101e⁺	+	406	= 609 leptons of 1 down quark
	-------		-------		-----	= ----- ---
	307e⁻	+	306e⁺	+	1226	= 1839 leptons of 3 quarks

Here are some of the most difficult questions in physics. I have answered some of these questions already, and I will answer the remainder of them later on in the book. I believe I have the complete detail and right answers concerning the particle physics as well as the secrets of the microcosmic and macrocosmic world of nature in our universe. All of them together lead to the finding of the ultimate being of our existence, which is part of the metaphysics.

1. How were all leptons (electrons, positrons and neutrinos) created?

2. How much is the accurate rest mass energy in electro-volts for each one of them?

3. What is a quizzical and phantom neutrino? How is it created? And how do you find it?

4. How are the quarks, protons and neutrons constructed?

5. What sets the electron's charge at $1.6 * 10^{19}$ coulombs? The negative charge "in" the electrons all repel each other, so what then holds the electrons together?

6. The muon has about 207 times the mass of an electron, thus the existence of muon raises another question. How can two equal particles alike in so many ways differ so greatly in mass?

7. What energy per second in electro-volts are the electrons producing being in different cells while whirling around their nuclei in the atoms?

8. Was any pre-existing mass in the universe? And if there was, what was this mass consisted of?

9. How did the creation start? How can cosmology or the structure of the universe and the laws, which underlie it, be explained and detailed?

10. Is this a finite universe perpetually pulsating and recycling in the eons of time, or is it expanding forever?

11. How are the strong and weak nuclear forces as well as the electro-magnetic and gravitational ones reduced to one only force, which is generating all others?

12. Is the gravitational force the cause that makes the earth orbit around the sun?

13. How the leptons (e^-, e^+ and ν) the blocks of the nature work all together in such a magnificent way? And how have they made a universe of infinity?

CHAPTER 7

THE EXTRA FRACTIONAL THIRDS OF CHARGES IN ALL THE ATOMS

THE SIMPLEST ATOM is the most common hydrogen isotope of protium. It consists of a nucleus neutrally charged and an electron spinning around it. The nucleus of a proton (P) contains 1836 leptons (306e⁻ and 306e⁺ and 1224v) or if I multiply 1836*3, I will get 5508 fractional charges of thirds. The 2754 positively charged (+) and the other 2754 negatively charged (-). This will indicate a nucleus in neutral status of the hot and cold powers (always remember that positive electrical charge means hot power and negative electrical charge means cold power). The orbiting electron (e⁻) as we know also contains 3 fractional charges, 2 of which are negative and 1 positive. So that in the whole family of this simplest and lightest atom there are all together 5511 fractional charges of thirds, (+) and (-). These fractional charges are 2755 of (+) or of hot power, and 2756 of (-) or of cold power. We are noticing here that in the whole atom there is an extra fractional charge of (-) or cold power. This extra fractional charge of 1/3 minus corresponds to the mass number of the same atom. And the more important thing is that, the number of the orbiting electrons will be equal to the number of the protons inside the nucleus of the atom, or equal to its atomic number.

If we look at the nucleus structure now carefully we will observe that it is consisted of 2754/5508 or 0.5 hot power and 2754/5508 or 0.5 cold power. This also could be expressed as positiveness and negativeness (+) and (-) of the nucleus. Now then the nucleus positiveness is 2754/5508=0.5. And its negativeness is 2754/5508=0.5. The orbiting electron positiveness is 1/3 or 0.3333333333 and its negativeness 0.6666666666. The difference in absolute thermodynamic temperature between the nucleus and the electron is: 0-(-1/3)=1/3, and it will develop a force of attraction. It should be clear that the intrinsic powers of cold-dynamics and thermodynamics in the nucleus are being counter acted as it has been mentioned before, so that a status of neutrality is in the nucleus itself. Any cosmic absolute temperature degree in the open space of the universe would not change or disturb these intrinsic powers in this atom or any other atom. Things are getting more complex in the depth of the atom domain, if I now take a deuterium atom of the hydrogen isotope. Its nucleus will be consisted of 1 proton (P) and 1 neutron (N), and therefore I will have a more massive nucleus. The neutron will contain 1839 leptons or 5517 fractional charges of thirds total. These fractional charges are 2758 of (+) and 2759 of (-). The proton on the other hand will contain 2754 of (+) and 2754 of (-), as mentioned before. Altogether in the nucleus of the deuterium atom will be 11025 fractional charges of 1/3. These fractional charges are 5512 of (+) and 5513 of (-). One extra fractional third of (-) in the nucleus itself. Also counting the 3 fractional thirds of the orbiting electron, I will have a total of 11028 fractional thirds, the two of them will be the extra negatively charged thirds in the whole family of the atom. One extra fractional third of (-) in the nucleus and one in the orbiting electron in the deuterium atom. The number (2) of the extra negatively charged thirds in the atom of deuterium corresponds to the mass number of the atom while the (1) orbiting electron (e⁻) (one again in this case) will correspond to the number of protons in the nucleus of this atom. The nucleus positiveness is equal to 5512/11025=.4999546485, while the nucleus negativeness is equal to 5513/11025=.5000453515, adding them together, will be equal to (1). The orbiting electron negativeness will be equal to 0.6666666666, and positiveness equal to 0.3333333333. The nucleus in this situation contains more thermodynamic power than the orbiting electron by the amount of: .4999546485 - .3333333333 = .1666213152 and less cold-dynamic power than that of the orbiting electron by the amount of: .5000453515-.6666666666=.1666213152. The more thermodynamic power in the nucleus and the less cold-dynamic power in it than that in the orbiting electron will amount to:

.1666213152 + .1666213152=.3332426304, almost the same amount of 1/3 difference in cold and hot power between nucleus and the electron.

I may now go further and I will examine the hydrogen isotope of the atom of tritium, with a single electron whirling around. Its nucleus of 1 proton (P) and 2 neutrons (N) have the sum of 5514 leptons altogether, (1836 + 1839 + 1839) or 16542 fractional charges of thirds of (+) and (-) plus 1 orbiting electron of 3 fractional thirds that will make a total of 16545 fractional charges of thirds altogether in the family of this atom. That is 8271 of (+) and 8274 of (-). In this case there will be 2 extra fractional thirds of

(-) in the nucleus and 1 extra fractional third of (-) in the orbiting electron. We will notice here that in the whole family of the tritium atom, there will be 3 extra thirds of (-), (or 3 thirds of cold power). This number of the 3 fractional charges of (-) in this atom will correspond to the mass number of the atom of tritium. And the 1 single electron spinning around will correspond again to the atomic number of the same atom, so that the number of the orbiting electrons and the number of protons in the nucleus are the same. The nucleus positiveness is: 8270/16542=.4999395478 and the nucleus negativeness is: 8272/16542=.5000604521; adding them both together will equal (1). The difference of thermodynamic power and cold-dynamic power between the nucleus and the electron intrinsically will be: .4999395478-.3333333333=.1666062145 and

.6666666666-.5000604521=.1666062145. By adding those 2 equal amounts of .1666062145 and .1666062145, I will have a difference of cold and hot power between the nucleus and the electron almost the same again, and that is .333212429.

If I follow the same procedure for any atom I will end up with the same results. Let us take the atom of helium (He), its nucleus consisting of 2 protons (P) and 2 neutrons (N), with 2 electrons (e⁻) in orbit. Or even the atom of actinium (Ac), atomic number 89 and atomic weight of 227.028. Almost similar or same situation will be illustrated. Starting with the (He) there will be 7350 leptons in its nucleus, or 22050 fractional charges of thirds altogether of (+) and (-). That is 11024 of (+) and 11024 of

(-), and 2 extra fractional charges of (-) in the nucleus. The whole family of the atom including the two electrons in orbit will contain 22056 fractional charges of thirds. These fractional charges are 11026 of (+) and 11026 of (-) plus 4 extra thirds of (-) negatively charged. We will notice here again that the 4 extra fractional thirds of (-) of the helium atom will correspond to the mass number, and the 2 electrons in orbit will correspond to the atomic number of the same atom, which is (2). The positiveness of the nucleus is: 11024/22050=.49995464852. The negativeness of the nucleus is: 11026/22050=.5000453514. The positiveness or the hot power of the two electrons in orbit is: 1/3 + 1/3=.6666666666. Their negativeness or their cold power is:

(-2/3)+(-2/3)=-1.3333333333. The more positiveness now between the nucleus and the two orbiting electrons is on the side of the electrons:

.6666666666-.49995464852 = .166712018. The more negativeness again is on the side of the 2 electrons in orbit. It is 1.3333333333-.5000453515=.8332879818. The clear negative electrical charge of the 2 electrons in orbit in the atom of (He) therefore is:

-.8332879818+.1667120182=-.6665759636. This negative charge divided by two (there are two electrons orbiting around the helium nucleus) will be equal to: -.33328798186 for each one of those two electrons. Almost the same difference between each one proton contained in the nucleus and each one orbiting electron. Finally the last example will be with one of the heaviest atoms in the periodic table of the elements, the actinium (Ac). Mass number 227 with 89 protons (P) and 138 neutrons (N), in its nucleus. Total leptons in its nucleus will be 417186. That is 1251558 total fractional charges of thirds in it, of (+) and (-) together. Or 625710 of (+) and 625710 of (-) counteracting each other, leaving 138 thirds of (-) extra fractional charges in its nucleus.

Also there are 89 orbiting electrons with 89 fractional thirds of (+) and 89 fractional thirds of (-) counterbalanced, and leaving the other 89 fractional thirds of (-) in orbit as extras. We are observing here that the total extra fractional charges of the minus thirds (-) are 138 in the nucleus of the actinium atom (Ac), and 89 in the orbiting electrons. Altogether 227 of (-) extra thirds corresponding to the mass number of the same atom. And the 89 orbiting (e⁻) are equal to the number of the protons inside the nucleus of the atom. The positiveness of the nucleus is: 625710/1251558= .4999448687. The negativeness of the nucleus is: 625848/1251558=-.5000551313. The positiveness or the hot power of the orbiting electrons again is 1/3 of all 89 orbiting electrons. It is 1/3*89=29.6666666667. The negativeness or the cold power of the orbiting electrons (which are 2*89=178) divided by 3 will be equal to:59.3333333333. The differences in the thermodynamic and cold-dynamic powers between the nucleus of the actinium atom and its orbiting electrons are as follows: the 89 orbiting electrons have more thermodynamic power than the nucleus itself, by the amount of: 29.6666666666-.4999448687=29.1667217979, and also more cold-dynamic power than the nucleus again by the amount of: 59.3333333333-.5000551313=58.833278202. That means that the total of the orbiting electrons (89) around the nucleus of the actinium atom, have a pure negativeness of cold-dynamic power, more than the nucleus by the amount of 58.833278202-29.1667217979=29.6665564041. And I say that this cold-dynamic power belongs to all 89 orbiting electrons. For each one of the electrons of course there will be a cold-dynamic power equal to: 29.6665564041/89=.333332094428. So that this fractional charge of almost 1/3 again (or cold power) of each one of the 89 orbiting electrons will correspond to each one of the 89 protons contained in the nucleus of the (Ac) atom. It is astonishing but I believe this is the reality of how the universe is constructed and developed. That should also be postulated. According to the particle physicists, the atoms may be lighter or heavier depending on the number of protons and neutrons being linked together in the nuclei. All atoms of the same elements have the same number of protons. The atomic number of natural elements ranges in succession from one to ninety-two. Also see the periodic table of the elements in the "The World Book Encyclopedia", pages 219-223.

CHAPTER 8

THE THREE KINDS OF HOT AND COLD TEMPERATURES IN THE UNIVERSE- HOW THEY APPEAR IN RATIOS AND INTENSITY

THE FIRST KIND OF TEMPERATURE

The number one is the temperature in the space, which shows how cold, or how hot something is. The top high of this temperature is about $5 * 10^9$ Celsius per lepton generated during the big bang. This top high temperature was condensed with the bottom low temperature. The ratio of the hot temperature was 13/27, and the cold temperature was 14/27. They existed this way in the pre-existing mass. The 13/27 ratio was responsible for expanding the universe. The idle cold temperature was responsible for contracting. As long as the universe was expanding this hot temperature was decreasing up to the zero Celsius degree and the volume of the space was increasing until it reached 1^0 Celsius. This kind of temperature has made the billions of stars and galaxies in the universe. While the universe was expanding there were serious changes in the cosmic times that could be graphically shown in a circle of a diagram.

I had discovered these ratios shown above when I was working on the construction of the quarks when I found out that two neutrinos had to be connected with one electron or one positron to construct a particle of trio. After this I concluded there should be a cosmic equation and this was one (two in one) that consumed the total number of leptons in the universe. See chapter 6. The extra cold of the above two ratios of 13/27 and 14/27 will make this universe pulsate and regenerate in the eons of time. Without these two ratios of 13/27 heat power (HP) and 14/27 cold power (CP) in one proportion, and without the leptons with their proportional fractional charges in thirds, no groups of three quarks could exist and no protons and neutrons would be in the universe. With no protons there could not be any nuclei for the microscopic planetary systems of the atoms, and without atoms, no stars and galaxies would be made.

THE SECOND KIND OF TEMPERATURE

This temperature is the top high of about $5 * 10^9$ Celsius and bottom low of

–273.16 consecutively contained intrinsically inside the leptons. These top high and bottom low temperatures condensed in an infinitesimal piece of a spherical mass of a lepton will change it either to an electron, positron or a neutrino. The electron will consist of two thirds of bottom low temperature and one third of top high temperature, all condensed in one tiny piece of a spherical mass. The positron will consist of two thirds of top high temperature and one third of bottom low temperature all condensed again in one tiny piece of a spherical mass. And finally the neutrino that will contain one and one half third of top high temperature, and one and one half third of bottom low temperature, all condensed in one tiny spherical piece of mass. These different quantities of top high and bottom low temperatures condensed into the total leptons in the universe existed in a proportion of 2/3v and 1/3 (e^- + e^+). This made them work in the atomic planetary systems for billions of years. This discovery of the leptons with their mathematical precision in fractional charges of thirds was in my opinion the biggest cosmological event of the millennium by a human on earth. The fractional charges of thirds in a microcosmic universe of 9 leptons will indicate and verify the ratios and the proportion of the initiated hot and cold temperature of 13/27 heat power HP and 14/27 cold power CP in the universe intrinsically condensed. See Figure No.7 at the end of this chapter.

THE THIRD KIND OF TEMPERATURE

The third kind of temperature is hidden in a neutron, which I believe it reflects a live universe. This neutron consists of 1839 leptons, which contain 5517 fractional charges. Of the 5517 fractional charges, 2758 are positively charged and 2759 are negatively charged. The ratios are 2758/5517 hot temperature of thermodynamic power and 2759/5517 cold temperature of cold-dynamic power. That means there is 1/5517 extra cold temperature in every 5517 parts of solid masses of an existing universe of stars and galaxies before it dies and becomes inactive at 0^0 Celsius and downwards. If you analyze the two proportions of 13/27 and 14/27 and also 2758/5517 and 2759/5517 you will find in the first there is 1/27 extra cold temperature in the universe either being dead or alive, and in the second there is 1/5517 extra cold temperature only when the universe is alive and active, where protons and neutrons are existing to compose the stars and galaxies. Subtracting 1/5517 from 1/27 I get: (1/27)-(1/5517) = .036855779. It is a significant difference between these two proportions. The 13/27 and 14/27 are acting the same way from the beginning of the creation until the universe dies when it reaches zero absolute temperature. But the ratios of 2758/5517 and 2759/5517 are acting only in a live universe. And that is why on earth we have an extra cold power of 1/5517 and not 1/27. The same thing is happening as the earth for all the planets, stars and galaxies in the universe.The humans could understand that we all exist in a very complex universe. But at the same time in a universe with its laws of the nature being similar and unchanged for all moving things, as Einstein had stated with different words. I would like to make a remark here, and say that if the initial proportion of the two ratios of the pre-existing mass was not such as 13/27 and 14/27, and was not to be kept throughout the universe evolution, no acting hot energetic masses of stars and galaxies would necessarily be developed in the space of our universe.

These above ratios were responsible for the creation and construction of the leptons by necessity. These leptons were the blocks of the NATURE. Their destiny was to build the universe. Condensing the two powers of hot and cold temperature in one entity made the leptons. These three kinds of powerful leptons, of male, female and neutral with their fractional charges of 1/3 in themselves, were running in the earliest cosmic time, one after the other, to get together and grow heavier to build the quarks, and the quarks made the neutral protons, and neutrons, or the nucleons. These in turn could make the heavier and heaviest masses of stars and galaxies with an extra cold power within themselves, equal to 1/5517. That meant that, with the protons and neutrons in a live universe, not only would they themselves have grown to huge bodies of mass like the planet earth, where we humans live on, but also they reduced the extra cold power from 1/27 generated during the big bang in the two ratios of 13/27 and 14/27 to 1/5517. You may compare an infinite volume of space with an extra cold power of 1/27 and 1/5517. In brief, without protons and neutrons in space, more serious cosmological problems could emerge. No hot energetic masses, no stars and galaxies or planets and other celestial bodies could ever exist. It would rather look like an empty volume of space!

Finally in my opinion, the ratios of 13/27 and 14/27 are responsible for:

1) The kind of leptons created in the universe during the big bang occurrence.

2) The proportional number of each one kind of leptons to the whole universe.

3) The construction of the fractional charges in each kind of all the leptons.

4) The number of fractional sections separating each lepton in all kinds of them.

5) The amount of charge in each section of any one and all kinds of leptons.

6) The overall resultant fractional charge of all the leptons needed to be connected in one particle, in order to represent and reflect one universe.

Fig No. 7

These nine leptons of two electrons, one positron and six neutrinos, shall constitute a microcosmic universe.

These leptons will have the ratios of 13/27 hot power, and 14/27 cold power.

CHAPTER 9

MASS, DENSITY, VOLUME, TEMPERATURE AND HEAT AND COLD POWER RATIOS

I MAY OBSERVE in a diagram of a circle, the whole cycle of the universe evolution with all its changes in mean densities, mean temperatures and volumes of space corresponding to the distance the universe has travelled in its own cycle. I will make a parenthesis here, and refer to the book "THE BIG BANG, The Creation and Evolution of the Universe" by Joseph Silk. On page 131 he says "Presently, the radiation temperature is only 3 degrees above absolute zero, and the density of the radiation amounts to approximately 1/10,000 of the total mass density that we can observe in the form of galaxies and stars. However at progressively earlier epochs, the radiation density played an increasingly greater role. The ratio of the density of radiation to the density of matter is proportional to the radiation temperature. Thus when the temperature was roughly 10,000 times higher than at present (or, when the universe was 1/10,000 of its present size), the radiation density was equal to the mass density."

He refers to the equation Density multiplied by the volume of a sphere is constant. Joseph Silk also said that density must be proportional to the inverse of the volume of the sphere or when (R), "the radius of the shell" approaches zero the density becomes arbitrarily high. There was never any doubt that he was right in every one of his conceptions above. Later I spent a long time studying the gas laws of physics. These were the Boyle's law, Charles' law and Avogadro's law in order to find and solve the mysteries of the densities at any point of the circle during a whole cycle of the universe's evolution, and eventually the mean absolute temperatures and the volumes of the space, which will correspond to these densities at any particular cosmic time. Aristotle said "Since the circle is a closed line and perfectly symmetrical it provided a symbol of what is unending and absolute harmonious."

Accepting that our universe has the shape of a sphere, it could be tolerable that the whole globe of our universe is made out by an infinity number of imaginary circles formed by orbiting masses in the planetary systems around the world. These circles could solve most or all of the mathematical problems in the whole universe by geometrical procedures, as this is one of the main topics of my cosmic theory. The main factors concerning our perpetually pulsating universe are: the total pre-existing mass (tM) of (13/27 HP) and (14/27 CP), the mean density (mDen), the volume of space (Vol), and the mean absolute temperature (mT). All these factors are related to each other. In every moment at any cosmic time in the space of the universe, there is a certain degree of mean temperature, a certain volume and a certain mean density. The densities, the temperatures and the volumes in the universe started to change from the very beginning of the big bang, in a zero degree of the circle, as Figure No.8 and 9 shows. From that point (zero degree of the circle), the top high temperature (it should be about $5*10^9$ degrees, for explanation see chapter 21), and the top high density of 100%, will start to be undergoing decreasing values, while the volume would be increasing. This will be continuous up to the point of the circle where the average temperature will come down to 1^0 Celsius in the space. Up to this point, the travelled distance by the universe should be 173.333333333^0 or 3.02523737 radians on the circle corresponding exactly to 13/27 ratio of the heat power in the pre-existing mass. From that point, and down to 180^0 of the circle or the zero degrees of Celsius there is still another 6.6666666667^0 or one half of 1/27 of the total contraction. When the temperature touches the 1^0 Celsius there is no change of (Vol) or (mDen). No expansion or contraction is taking place, as long as the temperature is steady at 1^0 Celsius. From 1^0 to 0^0 Celsius the temperature will still be falling down, but the (Vol) will start to decrease now slowly at first, and then progressively rapidly, especially when it is approaching the zero temperature Celsius or the 180^0 of the circle. On the other hand the (mDen), would start to increase slowly at the beginning and taking momentum close to 0^0 Celsius temperature. Let us not forget that (mDen) is proportional to the inverse of the volume in a sphere at any cosmic time. Although there is only 1^0 Celsius difference between one degree and zero temperature there is quite a traveling distance (Dist) by the universe of 6.6666666667^0 on the circle (see Figure No.8 and 9). That will show how fast and how far the universe is traveling to bring down the average temperature in the space by 1 only degree Celsius. From zero Celsius and lower to -1^0 Celsius the temperature would still be decreasing. The (mDen) will be increasing again, and the volume will be decreasing from 0^0 to -1^0 Celsius. The distance again from 0^0 to -1^0 Celsius on the circle diagram will be 6.6666666667^0. From -1^0 Celsius and down, the still decreasing temperature will force a further decrease in volume and increase in (mDen). From -1^0 Celsius, cold-dynamic power becomes higher and higher toward the end of the contraction of the universe, and at the same time the speed of the contraction will be decreasing progressively (see Figure No.3). In other words the factor of the absolute temperature would follow a decreasing value from the very beginning of the big bang to the very end of the cycle, but the volumes and densities would get different values and directions at different cosmic times. An important phenomenon will be observed when expansion of the universe comes to an end at 1^0 Celsius. It starts by the electrons in the oscillated atoms from $5*10^9$ Celsius and it is over when it comes down to 1^0 Celsius.

The distance from 1^0 to -1^0 is 13.3333333334^0 or 1/27 on the circumference of the circle. The temperature would never stop decreasing until the end of the cycle. Later you will find more details about it, and you will also learn how slow and how fast the universe is running its cycle at different cosmic times. The volume and the mean density will be changed at any cosmic time in such a way that when one of the factors is getting higher, the other will be getting lower; so that their product will remain constant forever. If for example at the start of the big bang, the volume of space was 1%, then at the end of its maximum expansion ending at 1^0 Celsius it would reach the value of 100%. Where as the (mDen) at the beginning of the big bang, where the mass is an intimate contact is at its maximum value of 100%, and at the end of the full size expansion the same (mDen) will come down to 1%. The mean density is always intrinsically consisted of both the 13/27 HP and 14/27 CP ratios of the pre-existing mass. The total mass though, will be influenced by the 13/27 HP ratio, which is the real changing power that makes the universe expand and contract, and continuously pulsate in the eons of time. This 13/27 HP is an active mass responsible for all of the changes in the universe at any cosmic time. These changes may be indicated as a percentage on the closed line of a circle diagram, or on a diagram of axes and coordinates. It is the main issue and the main cause of anything that has a motion in the whole universe. This active mass would release its total energy into the space of our universe. While the energy is released, the mean temperature, the mean density and the volume of space are changing. The amount of these changes and the duration of the time, for each one of them, will be dependent on the different cosmic times. This would last until the whole universe dies at 360^0 of the circle diagram in an absolute zero degree of temperature. I would recall that heat power means, "mass multiplied by mean temperature, and mass means volume multiplied by mean density." The 1/27 HP for example out of the 13/27 has a value of 3.70370370%. This is the same percentage that would also be indicated at the same time, to show how much the volume of the universe has expanded out of the total 100%. The 13/27 HP represents the 100%. The 13/27 HP plus the 14/27 CP will equal the total pre-existing mass. When the 13/27 HP ratio has been consumed somewhere at the closed line on the circle, the energy released from this heat power ceases to exist. There will be no more thermodynamic power in space, and there will be no more energetic temperature coming from the active mass. The (mT) reached the 1^0 Celsius, and the (mDen) the 1%. Confusing? I have no other way to explain them. See Figure No.8 and 9.

Figure No.8 deals with the changes of the (mT), the (mDen) and the (Vol) of the universe evolution, during the different cosmic times.

Fig No. 8

From 0 degrees to 45 degrees of the circle, the mass is in an intimate contact. There is a boiling temperature, the mean density and the mean temperature are decreasing, and the volume is increasing.

Zero degrees of the circle.
One second prior to the big bang, the temperature was -273.16 degrees Celsius before it exploded, and after the explosion, the temperature went up to $5*10^9$ per Lepton.

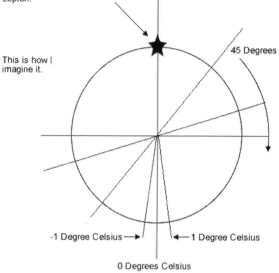

This is how I imagine it.

45 Degrees

-1 Degree Celsius ⟶   ⟵ 1 Degree Celsius

0 Degrees Celsius

From $5*10^9$ temperature up to 1 degree Celsius, or from zero degrees of the circle to 3.02523737 radians of the circle, the temperature and the mean density are decreasing, while the volume is increasing.

From 1 degree to 0 degrees Celsius, or 3.02523737 to 3.1415926536 radians of the circle, the mean temperature and volume are decreasing, and the mean density is increasing.

From 0 degree Celsius to -1 degree Celsius, or 3.1415926536 to 3.25794793706 radians of the circle, the mean temperature and the volume are decreasing while the mean density is increasing.

Finally, -1 degree Celsius to absolute 0 degrees Celsius, or 3.25794793706 to (2π) radians, the mean temperature and the volume are decreasing, while the mean density is increasing. The speed of the universe will slow down progressively towards the end of its life, or the end of its cycle.

Fig No. 9

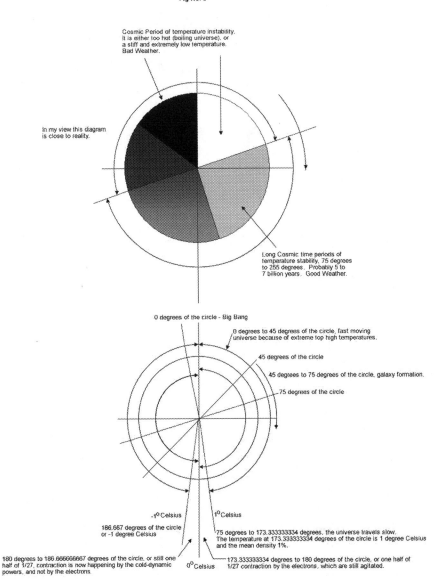

Cosmic Period of temperature instability.
It is either too hot (boiling universe), or
a stiff and extremely low temperature.
Bad Weather.

In my view this diagram
is close to reality.

Long Cosmic time periods of
temperature stability, 75 degrees
to 255 degrees. Probably 5 to
7 billion years. Good Weather.

0 degrees of the circle - Big Bang

0 degrees to 45 degrees of the circle, fast moving
universe because of extreme top high temperatures.

45 degrees of the circle

45 degrees to 75 degrees of the circle, galaxy formation.

75 degrees of the circle

-1° Celsius   1° Celsius

186.667 degrees of the circle
or -1 degree Celsius

75 degrees to 173.333333334 degrees, the universe travels slow.
The temperature at 173.333333334 degrees of the circle is 1 degree Celsius
and the mean density 1%.

180 degrees to 186.666666667 degrees of the circle, or still one
half of 1/27, contraction is now happening by the cold-dynamic
powers, and not by the electrons.

0° Celsius

173.333333334 degrees to 180 degrees of the circle, or one half of
1/27 contraction by the electrons, which are still agitated.

Fig No. 10

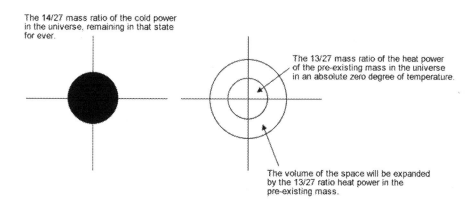

The 14/27 mass ratio of the cold power in the universe, remaining in that state for ever.

The 13/27 mass ratio of the heat power of the pre-existing mass in the universe in an absolute zero degree of temperature.

The volume of the space will be expanded by the 13/27 ratio heat power in the pre-existing mass.

Summing up, the point of the average temperature of 1^0 Celsius on the circle is very important concerning the thesis of our universe. At this point, the distance of the universe should correspond to 173.333333334^0, and there it will cease from expanding anymore. From this point it will start to contract fast by the exclusive work of the electrons in the atoms, still being oscillated from the thermodynamic powers. This is only until it reaches 180^0 of the circle, or the zero degrees Celsius. This distance is corresponding to 6.66666666667^0, or to one half of 1/27 of the total distance on the circumference of the circle diagram. The (mDen) from 1^0 to 0^0 Celsius will be increasing rapidly and the (Vol) following suite, decreasing too fast as well. At 0^0 Celsius the electrons will stop turning around their nuclei. They won't move anymore. They should be consolidated with the nuclei. Also from 1^0 to 0^0 Celsius the electrons are bringing down the temperature, and eliminating to zero the distance between themselves and the nuclei, thus an extensive decrease in the (Vol) of the atoms will occur.

Electrons and nuclei become a solid mass, and this is going to be reflected across the whole space of the universe in record cosmic time. From the 180^0 of the circle or 0^0 Celsius to -1^0 Celsius, which is corresponding to 6.66666666667^0, or to one half of 1/27 of the total distance again on the circumference of the circle, the stationary cold-dynamic powers will begin to work. From that point of -1^0 Celsius to the absolute zero there is another distance of 173.333333334^0 degrees of the circle to be travelled by the universe, and another 13/27 ratio for its contraction. The mean density will be rapidly increasing now, while the temperature and the volume will be decreasing. The universe will follow its destiny towards the end of the cycle. The (mDen) in the universe increases and the (Vol) decreases. The speed of the universe progressively will be getting slower especially when it is approaching the end of the cycle, in the last quarter of the circle (see Figure No.3). This all would be done again by necessity. Heat power and cold power ratios normally distributed on the circumference of the circle during the different cosmic times. But this is an intractable universe.

In my opinion there is not only one average temperature but also three of them. I recall that three in one were the light, the dark and the space. Three kinds were the leptons the blocks of the nature. Three kinds were the sub-atomic particles of trios that made the three kinds of quarks, each one of them being connected by 1/3 of a fractional charge and so on. Many other threes everywhere, the number three is present here too. This will have to remind us again of a symbolical representation of the TRINITY. The union of THREE PERSONS IN ONE GODHEAD very frequently mentioned by the Christians. In my opinion this TRINITY represents the father, the mother and the Holy Spirit (see chapter 4). I would say then that a) one average temperature is within the stars and galaxies, b) another one in the interstellar space, and c) the third one is in the borders of the global universe. When the two ones (a and b) are equal to the third (c) that would be the end of a live universe, with an absolute zero average temperature in it all over. In my view, I could not understand a total average temperature in the universe, without taking into account all the above mentioned three of them. It looks though that our universe has made its choice to dissolve itself piece by piece. That is, whole planetary systems of stars and galaxies are absorbed and vanished when they are approaching the outermost perimeters of the global universe. In this way the universe made up of billions of stars and galaxies would come to an end when the time comes.

These stars and galaxies will start from the center of the globe, and progressively they will change their circular directions to elliptical and spiral ones, and after a few billions of years, when they will be approaching the outer perimeters of the globe, they will be thinned out and weakened in power, to be easily attracted and absorbed by the strength of the concentrated cold-dynamic powers there (see Figure No.13).

Re-examining Figure No.8 and 9, you may remark that the total expansion and contraction for one cycle of the universe, is divided in three parts again:

a) The distance of the expansion period shown in the circumference of the circle by the thermodynamic powers equal to 173.333333334^0,

b) The distance of the contraction period by the still agitated electrons in their atoms of 6.66666666667^0 taking place again above zero Celsius in the thermodynamic side of the circle, plus the same distance of a contraction period now taking place below zero Celsius by the cold-dynamic powers (see Figure No.18), and

c) The rest of the contraction for the total cycle by the stationary cold-dynamic power of 173.333333334^0. If I add them all I would get the 360 degrees of the circle diagram. In radians, it should be:

$3.02523737 + (.116355283466 + .116355283466) + 3.02523737 = 2\pi$. In ratios it is:

13/27 + (0.5 * 1/27) + (0.5 * 1/27) + 13/27 = 1. This is correct.

But if the pre-existing mass was (A) and the (mT) per lepton was 5 * 10^9, when the big bang occurred, the thermodynamic power should be a ratio of 13/27 * (A) heat power to complete a total expansion of the universe. For the contraction period to complete the cycle, there should be an opposite cold-dynamic power of a higher magnitude as that of the thermodynamic one, to bring the universe to where it has started. It should be (A) * (14/27).

One should remark here that the ratio of (14/27) would not be equal to the ratio of (13/27). Of course there should be a 1/27 extra cold power of the total pre-existing mass in order to have a moving and pulsating universe in the eons of time. The hot and positively charged power is the 13/27 ratio of the pre-existing mass. The cold and negatively charged power is the 14/27 ratio of the same pre-existing mass.

Now then, how would this whole thing turn out? I thought about it and I found out that if these minus degrees of –273.16 from zero Celsius to the absolute zero were raised to the 4th power, it could give me a temperature number of contraction power totaling

-14/27 of the pre-existing mass. Indeed -273.16^4 = (-5,567,605,000) and in my opinion it was the right answer. If I accept that the top high temperature when the big bang occurred was 5 * 10^9 as I believe and I have a full explanation for this matter. This number was pretty close to the minus temperature which multiplied by the total pre-existing mass, in a state of absolute zero degree, would give a -14/27 cold-dynamic power. COULD THAT MEAN THAT THE COSMIC TIME OF A RETARTED CONTRACTION PERIOD OF OUR UNIVERSE WOULD BE THE SAME AS THAT OF THE EXPANSION ONE? It may be so. Adding both numbers of the thermodynamic top high temperature and the bottom line cold-dynamic one, I have the number 10,567,605,000. If this number was equally distributed to the circle diagram, eventually it would give 5,088,106,111 thermodynamic degrees for the (13/27) of the expansion and then between 1^0 and –1^0 Celsius 195,696,389 thermodynamic degrees for the .5 * (1/27) of the contraction by the electrons in the positive side of the circle, and 195,696,389 cold-dynamic degrees for the .5 * (1/27) of the contraction in the negative side of the circle (see Figure No.18 for more details). Finally another 5,088,106,111 cold-dynamic degrees for the rest of the contraction to end the cycle on the circle diagram. The total numbers of temperature to produce the hot and cold powers are 5,088,106,111.11 + 195,696,389 + 195,696,389 + 5,088,106,111.11 = 10,567,605,000 corresponding to 13/27 HP and 14/27 CP. 5,567,605,000 for the cold power and 5,000,000,000 for the hot power.

The cold-dynamic powers don't move either up or down as the thermodynamic ones. They are stationary and condensed with the hot temperature in one entity in the eons of time. Demonstrated both these powers as they are functioning, are seen on page 55, Figure No.10. Without the existence of the atoms, I think no external planetary systems would ever appear, and the volume of space would be expanded and contracted automatically for ever. At present the huge hot and cold powers make the universe run fast or slow in different cosmic times. But still, the stars and galaxies in the space seriously reduce the speed of the universe expansion. This is a cosmological event and it is happening because the released energy by radiation is an amount of light, locally rotated. It is emitted by the nuclei and around the nuclei, within the planetary systems themselves. It is done in such a way, that only the minimum radiation in light or temperature could be defused to the surrounding area, or the interstellar space. If the universe takes $2(2\pi)$ billion years for one cycle, the long duration in cosmic time of temperature stability would probably last in my opinion about 6 billion years. The traveling distance by the universe would be between 75^0 and 255^0 of the circle. I would imagine that a diameter cutting off the circumference of the circle at the right spots (see Figure No.9), would divide the cycle into two equal parts. One part of good weather and stability in temperature (it should be like a sunshine), and the other part of instability in temperature and bad weather (too cold or too hot). I will dare to still go further and presume that vegetation, animal and human life made their appearance at the beginning of good weather and probably they were bigger in size and lived longer periods of time in our cosmos. But where are we now? Is our universe close to its end or not? Did it touch down the average temperature of zero Celsius? And if it did, how far down below are we at the present time? Is it close to -273.16^0 degrees of Celsius? I am seriously concerned about this matter. In fact Joseph Silk on page 131 in his book "The Big Bang" says: "the average temperature in the universe now is 3 degrees above absolute zero." What average temperature did he mean? Is that an average temperature of the interstellar space? Our solar system is within the Milky Way galaxy full of planetary systems still in warm weather. But it may be that our solar system is spiraling away from its center in to the open space. If our galaxy is weakening, it might be a sign that we are close to the outer borders of our globe. No bells or sirens should start to go off for our humanity yet. But there is not much cosmic time left, before we are absorbed deep into the dark of a metaphysical world. Our universe will not last forever.

CHAPTER 10

NEW IDEAS IN COSMIC PHYSICS

BATTERIES IN THE SKIES

The planetary systems in the skies involving two or more objects of mass will demonstrate the role of electric batteries with a certain capacity. The capacity of these objects is their heat power, which is measured by the amount of their total mass multiplied by their mean absolute temperature degree per lepton. The more active body of mass with the higher heat power constitutes the positive pole of the electric battery. As an example the heat power (HP) of the sun determines the number of amperes which could be produced out of the total capacity of its heat power consumed in time (t), which time (t) will be depended on the voltage developed or the difference in heat power between the sun and any of its particular planets, or between the sun and all of its planets simultaneously. Let me change the picture for one moment in the universe, for instance from objects of mass releasing energy, into a battery producing electric current in amperes per second. A huge total mass of our sun combined with its 7 to $8,000,000^0$ Celsius per lepton. (The Encyclopedia Americana 1997 edition on page 12 refers to about 10 million degrees Celsius in the center of the sun). That will make it very energetic and a high activity body of mass and of course this furnace of a thermonuclear reaction will become a positive pole of a battery to release huge amounts of energy, perhaps billions of watts per second on the earth's surface in the form of a speed of light emission or radiation in the square.

This will become a closed circuit of an electric current because the whole action is performed in a stage of a closed local balanced unit. The negative pole will be a particular planet whichever planet that may be, or all planets together around the sun. The sun with its huge power (HP) will be able to form units of electrical batteries with each one of the weaker negative poles of the planets independently. The same performance is played with the atoms and their electrons whirling around the nuclei. Every electron negatively charged becomes a prisoner from a corresponding proton in the nucleus of the atom, which is playing the role of our SUN in a microscopic way. The battery's positive pole connected in the space with a negative pole of an object found somewhere in the distance is almost perfect. It is an almost closed electric circuit. I say almost, as sometimes some of the electric energy can be lost in the form of light and radiation emission through the interstellar space. The same thing exactly is happening in any kind of earthly machine producing energy. So that the strength of the positive current of sun will be flowing in almost straight lines from the center pole of the sun, to the center pole of the earth (which is taken as an example), or the center pole of the earth to the center pole of the moon and so on. See Figure No.11 on the next page.

However, because the earth and the moon constitute another independent small battery, these two objects as one unit would constitute one negative pole so that the sun will act on the earth and the moon as if they were a single body with its center of course now being about 1600 kilometers below the earth's surface, and this spot is the earth-moon barycenter. Now as a result the negative masses including the whole mass of the earth and the moon will be pulled around the sun in an elliptical orbit away from the central line of the positive current of a very hot vacuum created between the sun and the earth-moon together. There is only one exception. Because the moon is pulled by both the earth and the sun, the earth will be following a wobbly path on its elliptical track around the sun. The duration of this battery connection and energy production will probably last billions of years of life, depending on the capacity of the sun's heat power and on the difference in heat power between positive (+) and negative (-) poles and the magnitude of the voltage created and the loss of energy to the interstellar space! Nevertheless, because the negatively charged planets around the sun will convert the positive current of the sun to a kinetic energy of negative current, they will at the same time resist the positive force of attraction of the sun, and because all this action is played in the same closed electric circuit of a thermal local equilibrium probably the minimum ever amount of the electric energy of the battery will be lost to the open space of the universe. Figure No.11 illustrated on the next page will give the reader a better understanding of how the objects of mass in the space play the role of an electrical battery.

Fig No. 11

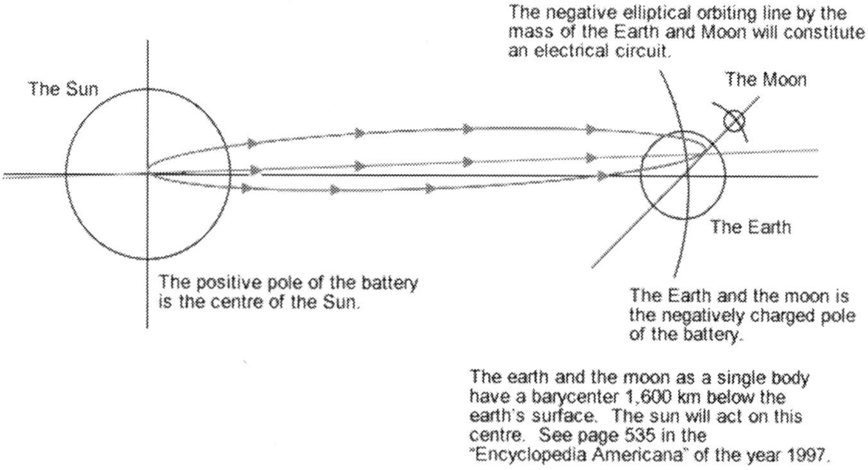

The negative elliptical orbiting line by the
mass of the Earth and Moon will constitute
an electrical circuit.

The Sun

The Moon

The Earth

The positive pole of the battery
is the centre of the Sun.

The Earth and the moon is
the negatively charged pole
of the battery.

The earth and the moon as a single body
have a barycenter 1,600 km below the
earth's surface. The sun will act on this
centre. See page 535 in the
"Encyclopedia Americana" of the year 1997.

OUR SOLAR SYSTEM WITH ITS PLANETS IS COMPARED TO AN ATOM

Let us assume now that our solar planetary system gets disturbed. Earth for instance, happens to be pulled away by some other powerful heat power source. The sun with the rest of its planets will behave like an ion looked upon as a positively charged one. The whole solar system will become unstable, changing the previous status of a perfect unit of a local thermal equilibrium. It will be swinging into space looking forward to catch up with another negative object of mass similar to that of the missing earth and attract it to its family in order to stabilize its thermal local equilibrium. The sun being in this condition will rather radiate more of its heat power energy into the space than it was doing before. This situation will get worse if the whole solar system moved away towards a cooler region of space. If it happens to get closer to a more powerful and very much superior heat power source, like an aggregation of stars or galaxies, the solar system would probably be pulled out towards this direction, being now negatively charged in relation to the source which would have a higher thermodynamic power than that of the solar system itself.

Every motion in the space of the universe is meant only when it is circular, elliptical or spiral. No particle or object of mass could stay completely idle in the space of the universe. It would always be a prisoner to another object of mass, of a superior heat power source as long as the universe is alive. Any object or particle of mass close to the extreme outer perimeter of the globular space will be sucked away and wiped out by the abyss of a crystallized cold-dynamic power in an absolute zero degree of temperature. In my opinion no strong beams of light, stars or galaxies, are stronger in order to defeat or avoid this hard and tremendous solidity of the cold-dynamics and go further away from that utmost last perimeter of the globular universe. All bodies of mass in space would have the same luck. There is no more space beyond the outermost perimeter of the globular universe. I believe that those black holes very often mentioned by the astronomers, are spots in space, which probably have an unimaginable depth and they are communicating with the Hades and ultimate grounds of death and metaphysics.

HOW THE EXPANSION AND THE CONTRACTION OF THE UNIVERSE INFLUENCE THE MISLEADING WORLD OF NATURE

If the whole universe is compared to a balloon it won't be inflated forever. Its globe will start to reverse its upward path of inflation to a downward path of deflation at some cosmic time. As long as the globe of the universe is expanding by a power rate emission of light, this emission must be coming from a heat power source of the prevailing thermodynamic powers in the universe. When the universe starts to deflate the cold-dynamic powers will be prevailing and as a result they will take over the contraction procedure. See Figure No.12. It will start from a time (t) when the big bang will occur, and it will be extended up to the point of the time (t'). From this point it will reverse the path until the point of the time (t"), where it will be the end of the universe cycle. The period from (t) to (t') is the period of expansion of our universe. And the period from (t') to (t") is the period of contraction.

These changes of directions and the period of the time taken in each one of them are important because they are going to indicate how the life of the misleading world of nature is influenced during these periods of expansion and contraction. I believe that these changes may influence almost everything in our world of nature. The upward course for instance, when the universe is growing up, the life of every object of mass could grow bigger and slower depending on the distance the universe has travelled on its cycle of the circle diagram. This case of extending a potential bigger and longer life could include the humans on our planet, the animals and the vegetation, while on the opposite direction of a contractible universe; life could become shorter and weaker including everything again on our planet. And to further detail this case of the universe, following the direction of night-day-night which is the (-), (+), (-); or dark-light-dark, I dare to infer that if creatures existed millions of years ago then they were of different building size and strength. They probably were living much longer than we are living today. That means that if the universe is in the process of contracting itself now as it is my belief, the more the time goes by, the shorter the life will become for humans, animals and vegetation. That could be so in my opinion, but only if we were left to grow up naturally of course, without medicine and medical care.

Now again in my opinion, humans have been developed with some certain degrees of absolute temperature and by the help of certain limited or unlimited and delicate combination of different atoms of a heavier or a lighter construction. The humans move, think and do things by their own power (self conducting). The how and where and what atoms are combined in the organs of the human body should lead to the inference of how we sometimes may be able to cure the illnesses or improve the health and the bad functioning organs of our bodies. Our human construction is probably made of by heavier or lighter atoms to produce a heavier or lighter human body. During the inverse procedure of the universe, (expansion turning to contraction) the other serious situation, which will be encountered, will be of a very high concern to the humans of our planet. This will be the procedure of the universe to advance progressively deeper into the night. The power rate of light emission, the electromotive forces and the forces of attraction will get weaker to a significant degree as the space will be getting cooler and cooler. This will happen as a result of the increasing stiffness and growing density of pressed masses in rigid cold interstellar spaces. This is a clear sign, that because of the rigid cold, that is why there is an increase in the resistance to the weakening power rates of light emission from the still remaining little heat power sources of thermodynamic powers in the center of the globe.

Fig No. 12

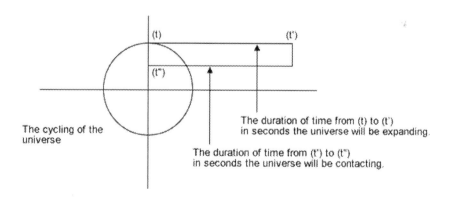

THE IMAGE OF THE UNIVERSE; DETAILS OF GLOBAL CIRCULAR, ELLIPTICAL AND SPIRAL MOTION IN THE SPACE OF THE UNIVERSE

A thoroughly comprehensive universe is the goal of the cosmic theory of thermodynamics. We have a good understanding up to now of a perpetual pulsating universe, but let us talk about it a little more in general. I believe that the image of the universe is a round globe like the sun and the earth in our solar system and all other celestial bodies in the universe as well as all the elementary particles of leptons, nuclei, atoms, etc. In this globe all the objects of mass in general are giving off their energy to expand the globe of the universe up to the point where it can no longer go any further. This point will be attained when no more expansion will be taking place any more in the whole universe. Most of the work in the energetic universe is being done by the stars and the galaxies, all of them while radiating their energy simultaneously are moving and spiraling towards the direction of the outer perimeter of the globe in the universe to their own destiny, attracted by the more prevailing dynamic coldness and darkness of the pre-existing mass in an absolute zero temperature which is the "Mother of Nature".

This globe of our universe (Figure No.13) shows that the most luminous spaces are around the cluster of galaxies. All these galaxies are shaping planetary systems and at the same time revolving the globe of the universe one after the other. They are following a spiral direction towards the outer perimeter of the invisible wall of our universe. That invisible wall is an impenetrable and crystallized wall of an absolute zero degree. In my opinion this general image of the evolutionary universe has been the same since the beginning of the big bang genesis. The first and the second law of thermodynamics say that " Internal energy of a system is introduced" and "The entropy of an isolated system tends to increase to the maximum value obtainable." (See the Encyclopedia Americana, 1997 edition, pages 347-348). On page 347 it also says "Energy is neither created nor destroyed." I think that the meaning of the above mentioned is that energy in the universe is always the same. It can neither be increased nor decreased. This law of the thermodynamics or rather better yet, of the thermo/cold-dynamics, in my opinion is a corner stone of our cosmology. This is true and it shows that there should be a finite universe existing. It is indeed a closed universe with limited borders. Its size is determined by the strength of the big bang, the ratio of the heat and cold power and the total pre-existing amount of mass. These are the factors of course, which will determine how much energy totally will be released in our universe from the beginning to the end, and how much the universe will expand and contract. The universe contains the trinity of light, dark and space in itself. There should not be another universe.

The veil of darkness in the extreme borders of the outer space somehow keeps the perpetual pulsating universe "suppressed inside" and it only allows it to inflate and deflate. That veil of darkness is the invisible dark mass in an absolute zero degree of temperature. It is a temperature of hot and cold condensed intrinsically, THAT I BELIEVE WILL CREATE THE MASS, NOT THE MATTER. This temperature of how cold or how hot something is, CREATES OUR UNIVERSE! This universe is the earth, the sun, the galaxies, the volume of space, etc. The duality of hot and cold of this temperature will also create the STRONG AND WEAK NUCLEAR FORCE, and THE ELECTRO-MAGNETIC and GRAVITATIONAL

ONES. A light bulb switched on and off, depicts the globe of the universe. It is like a balloon inflated and deflated. There is nothing existing beyond the veil of the global universe of course, except a converted crystallized lifeless mass like it has been explained before. Later the same veil of the crystallized coldness and lifeless mass will be the cause that will start the shrinking and capturing of the emptiness in the spaces remaining in the global universe, squeezing the spaces remaining more and taking the form of a sphere in an infinite density and solidity. An instant after a heavenly almighty big bang will recreate the same universe. It will repeat the work of releasing the energy of the imperishable and inheritable thermo/cold-dynamic temperature. It is a mysterious universe in a multiple magic image. This image gives us the impression of a light bulb, being the only thermic source in a cold empty space of darkness, with its incandescent element of tungsten burning in the center. We may see that progressively this light will be getting darker and cooler as it will travel through the space, it will meet the resistance of darkness and coldness to stop its forward maximum expansion in space.

The dynamic power of the light will allow it how far it will travel in a radial and circular direction. Curiously enough the same thing is happening with the whole universe. The finite universe! There is an end to it. The universe is indeed a light bulb. From the most luminous central volume of space, to the far end of the outer perimeter of the globe, there are shaping existing layers of less luminous spaces, colder, denser and darker volumes of spaces. The end of it is an abyss of an invisible blackness, of a Hades, which is described as "The place or state of departed souls, the world of spirits" on page 331 of the "Consolidated Webster Comprehensive Encyclopedic Dictionary". I believe that this Hades is a dark mass of condensed thermo/cold-dynamic temperatures, in a state of crystalline and absolute solidity of density, not going beyond. It is hard to seize the conception of this all and bring it deeply enough into your imagination, but that is what it is. When this globe is entirely deflated like a balloon losing all of its air, it will become a lifeless mass of a black sphere of solidity, of nothing else than a compressed mass of a more cold-dynamic power and less thermodynamic one.

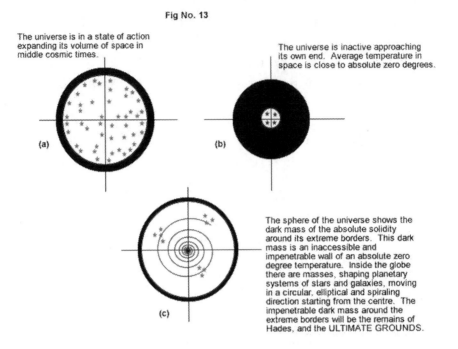

Fig No. 13

The universe is in a state of action expanding its volume of space in middle cosmic times.

(a)

The universe is inactive approaching its own end. Average temperature in space is close to absolute zero degrees.

(b)

(c)

The sphere of the universe shows the dark mass of the absolute solidity around its extreme borders. This dark mass is an inaccessible and impenetrable wall of an absolute zero degree temperature. Inside the globe there are masses, shaping planetary systems of stars and galaxies, moving in a circular, elliptical and spiraling direction starting from the centre. The impenetrable dark mass around the extreme borders will be the remains of Hades, and the ULTIMATE GROUNDS.

I would like to make you less confused and rather pleased to understand more of the details concerning the secrets of our universe. What is really happening to it is that a gigantic planetary system in our space is in action. That is, a sea of stars and galaxies we see at night in the skies are revolving around their axis, turning one after the other and at the same time all of them are circling around the global universe. From another point of view it looks like one planetary system is governing the whole space of our universe. This body of a huge mass of cluster, of galaxies and other objects of mass, as the cosmic time goes by is pulled out slowly and progressively towards the invisible wall of the extreme perimeters of the globe. This action is taking place because of a mutual force of attraction between thermodynamic and cold-dynamic powers on two opposite sides. One is the side of the cluster of galaxies and the other is the kingdom of the abyss and metaphysics, in the utmost perimeters of the globe. All that is happening there is an electromotive force developed between the two parties involved. These actions are exercised in all micro-cosmic and macro-cosmic planetary systems in space and all of them will constitute a gigantic system of masses pulled around slowly as one unit, towards the invisible dark mass of their destiny.

Their journey has started at the beginning of the big bang and it will end up at the extreme borders of the globular universe. These masses have changed or diverged their orbiting directions in different cosmic times from circular at the earliest epochs, to an elliptical one at the middle age and a spiral one as it approaches the end of the universe. The recession of some light orbiting galaxies being observed close to the invisible end of the global universe, is due to a strong force of attraction developed there between the light galaxies and the strong cold-dynamic power at the end of the globe stimulating the spiral motion of the galaxies that makes them fly apart, accelerating their speed to the end of the globe. It is very natural that these masses of galaxies as the time goes by, progressively they are developing a higher (EMF), with the approaching dark mass of absolute zero at the extreme perimeters of the sphere. This mutual force of attraction between these two thermo/cold-dynamic powers is becoming inversely proportional to the square of the distance in meters between the galaxies and the wall of the invisible dark mass. But make no mistake, forces of mutual attraction exist in an infinite number of planetary systems in space, and all of them compose the stars and galaxies, and all the other masses around the globe. All of them are slowly dragged by the extra 1/27 cold-dynamic power like one body of mass, until they meet their own end.

All my theory of cosmic physics including this chapter, in my opinion will be very important to our humanity. The idea of how the universe has been created and what was it that made it look like a materialized universe, I think has been a problem for our humanity in the past. Naturally the mystery of our cosmology has probably puzzled some theorists, physicists or astronomers. The particle physicists might wonder why the nuclear force is acting only in extremely small distances and why the electron has an electric charge of $1.6 * 10^{19}$, and many other important phenomena in space that they have no answers for. As an example I will mention that sometimes in the news I see things like "the mass and gravitational forces of stars and galaxies is not sufficient to keep them from flying apart, and that some hugely massive and invisible dark matter, must generate gravity to hold the universe." I think that much of a detailed explanation for this and many other issues is that we are dealing with a theory of classic physics. In my view the serious problem always is that we should find another theory, which should go beyond standard model. Before closing this chapter, I would also like to say that no motion in the space of the universe could ever be a straight line; not even the beams of light of radiation and expansion like exactly Einstein's example when he discovered and approved that a beam of light was curved while traveling in space and passing close by to a celestial body of mass. Beams of light in space look like they have a straight direction but as they travel in space; they are either more or less inclined depending on how far a body of mass is from their direction. If there is no celestial body of mass met, then the beam or the beams of light in my opinion should later follow a spiral direction and take the same curved line as that of the approaching utmost periphery of the borders of the sphere in the universe inside the globular space. It is a must that all the masses of the universe, either being stars, galaxies or plain beams of light are going to follow the same road and the same direction of their destination. In the end all of them will become prisoners of the outer perimeters of the globe, where there are the massive dark masses of the metaphysical world and the absolute zero degree. The whole universe is living in circles. All the movements of any kind of particle or object of mass must follow the globular shaped universe locally and distantly. Figure No.13 on page 66, part (a), (b) and (c) clearly illustrates the sphere of the universe and the directions of all circular, elliptical and spiral motions for all the masses inside the globe including the stars and galaxies.

CHAPTER 11

THE EXISTENCE OF TWO KINDS OF PLANETARY SYSTEMS IN THE UNIVERSE-RELATION OF LIGHT, TIME AND SPEED OF LIGHT

THERE ARE TWO kinds of **PLANETARY SYSTEMS IN THE WORLD.** The microcosmic planetary system (which will comprise all the atoms being the families of a misleading world of nature), and the macrocosmic ones of the stars and galaxies in the open space of the universe. Both these planetary systems occupy the smaller portion of the globe in the space of a live universe. These systems are functioning in a different way under different circumstances. In the macrocosmic external planetary systems for example, the stars and galaxies like our solar system, are in a nuclear reaction and huge temperatures are unfolded from inside of them that are millions of degrees in Celsius. These temperatures are not the same between the stars and the planets or even between the smaller and greater galaxies. These circumstances will cause high potential differences between themselves and very high electromotive forces, which in turn will develop high forces of attraction ending up with the form of planetary systems of small and large sizes in the open space. So that high potential difference of heat power among the stars and galaxies in the open space will necessitate electro-motive forces followed by forces of attraction and shaping of planetary systems of local thermal equilibrium.

This is one kind of planetary system in space. Situations in the atoms are changing drastically concerning their way and manner of how to develop and form their own planetary systems. Here the atoms do not need high external temperatures in order to develop potential differences between the electrons and the nuclei. They only need to have a temperature of a few degrees above zero Celsius in order to create high potential differences between the electrons and the nuclei. This is because they are helped by their intrinsic condensed highest and lowest thermodynamic and cold-dynamic powers in the universe. (See chapter number 6 about leptons). I will also have more details about it in the next pages. By necessity they will use their intrinsic hot and cold temperature to make them develop forces of attraction of high magnitude and form their famous planetary systems inside the masses of stars and galaxies in our live universe. This will be the second kind of planetary system in space.

THE LIGHT, TIME AND THE SPEED OF LIGHT

In my view the light, the time and the speed of light go together. They are displaying the image of the trinity (see chapter 4). There is no time and speed of light without light, or time and light without speed of light. The speed of light in the universe should be measured between the point that it begins to the point that it ends. The light in a circular motion travels the distance between these two points inside the atoms. If there is a long distance between the two points then the beams of light will follow a curved direction like that of the circumference of our global universe; but before they reach that point they might be deflected when in their path they meet another body of mass. The planet earth turning around the sun and the electron spinning around the nucleus are also playing another important role of light in space. The electron with its infinitesimal mass and high speed in revolutions will hold back the speed of light from the nucleus and stop it from spreading in space. This speed of light will automatically be converted to a speed of revolutions per second from the electron. The same situation will be observed between the stars and the planets or even between the cluster of galaxies; they will also hold back the speed of light from the stars or from the superior heat power sources of other galaxies in the huge external planetary systems, and it will be converted to a number of revolutions. Some of the light power is going to be lost in the process of going through and expelled into the interstellar space. It will be crossing the space losing energy to the square in meters of its traveling distance.

The light is always originating from a heat power source starting from 0^0 Celsius and up to about $5 * 10^9$ degrees Celsius per lepton. Under the 0^0 degree Celsius any existing light in space will start to get absorbed. No more radiation and expansion. I believe that the speed of light is slowing down proportionally to the square of its travelled distance in meters. Calculated by the physicists, it is about 299279 kilometers per second. In my opinion that speed was measured as soon as it emerged from its heat power source. No matter what the strength of the heat power source that has generated this light, it will have the same speed but its distance to be travelled will be different from other speeds of light emerging from different sources of heat power. The light will get a degree of power and a corresponding color and/or wavelength depending on its original heat power source. Its energy and traveling distance will be increased or decreased and always limited. It is like what was said in chapter 3. In my opinion the reddening light in space is not only coming from recession velocities of distant stars or galaxies, as scientists and astronomers might say; it could also be a light from a weaker source of the thermodynamic power of distant stars or galaxies.

As an example, if you try to set on fire a very small piece of paper in the middle of a dark room, you will have trouble finding the door and getting out of the room, and at the same time not being able to see how big the room is. But if you lit a bunch of papers which consist of more mass and consequently more heat power, you will be able to see a brighter room and the light will even travel further through an open door or an open window. In my view the basic principle for calculating the speed of light is to divide the distance the light has travelled from its starting point until it is totally wiped out by the time in seconds. Short distance from the source, the light maintains its initial speed of 300,000,000 meters per second (page 44 from the book "The Big Bang- The creation and evolution of the universe" by Joseph Silk).

In greater distances the light will get weaker and reddish in color, traveling at a lower rate of speed (because of the density in space) and tending to an elliptical and/or spiral speed. The bottom line is that sooner or later the light will be wiped out by the cold-dynamic powers. That is why the force of attraction in short distances is very powerful as you may observe this phenomenon in the depth of the atom where the nucleus pulls the electron with such a force that it will spin around with billions of revolutions in a millionth of a second. I think that the power of a light is equal to the thermodynamic kinetic force of a heat power source of an object of mass that has developed it, divided by the square of the distance in meters the light has travelled already up to that point where it is to be measured. Billions and billions of different lights are traveling in space from different heat power sources. They cross each other but they do not necessarily have the same energetic power. Galaxies also are spinning around their axis while at the same time they are traveling around the globular universe; and here is another phenomenon in the skies. Reading magazines of astronomy and newspapers, I found out that sometimes galaxies long distances away are instantly disappearing. There are two ways that this can happen. I believe one is that galaxies are orbiting the globe and spiraling away from the center where they were before, and while they go around they get lost and disappear. But they might be seen from the other side of the globe like new galaxies this time! The other case is when the galaxies are too close to the invisible wall of Hades and metaphysics. In this case they might be instantly swallowed down by the most powerful cold-dynamics to become a pre-existing mass.

CHAPTER 12

SUMMARY OF THE CREATION AND THE DEVELOPMENT OF OUR UNIVERSE-THE LIGHTER AND THE HEAVIER FORM

THE BIG BANG brought forth the light of life and the presence of the leptons with their fractional charge, which developed the electromotive forces and in turn were responsible for all developments afterward of our universe, etc. An extremely critical cosmic time in our universe had come at the moment when the neutral quarks stopped growing heavier. Why didn't these quarks get any heavier? Why didn't they become any lighter in weight instead? Why did they always remain enigmatic quarks, and they have never been taken apart on earth? I would say that each one of these neutral quarks of 612 leptons dominates the issue in a further process of building up the whole universe.

All the particles of leptons, trios, quarks, protons and neutrons seemed to have a pre-existing "know" of how to do it all perfectly. They were all created and developed in a few critical moments; and they had an order of automatic procedure to construct and produce themselves in a row during the first stage of the light part of creation in our universe. And of course taking the advantage of the highest cosmic heat power combined with the cold power condensed intrinsically in these particles, they gradually were able to build up a universe to pulsate for billions of years. There still might be though some problems for physics to be cleared up and solved. The conception and the right direction of this cosmic theory in my opinion, leaves little doubt that all these and other problems in general are under way to be cleared out.

I conclude then that up to now we have learned quite a few things but not all of them are explained in detail, and of course there are other problems to be solved by the science of mathematics. We did learn the leptons creation and construction with their over all proportions of each one of the three kinds to the whole universe, along with the construction of the two main particles of the nucleons (proton and neutron) which are to build a whole world of nature with planetary systems, stars and galaxies. We have also learned at the same time how the "COSMOS" and its root were created, the origin of creation of the world, and the doctrine of the origin of formation of the universe. That is our "COSMOGONY". After the heavier form of the theory comes the structure of the universe and the laws which underlie it. Scientists say that mathematical symmetry means that we can change our equations in such a way that the initial and final forms describe the same physics. What kind of symmetry is there that possesses physical forces? In my opinion this is a mathematical equation, which will start from (0), it will go to (A), then from (A) to infinity (∞), and from infinity back to (0) where it started! (A) represents any kind of number, mass, light, temperature, etc. Newton (1642-1727) has said: "Universe is having a logical structure which would be properly interpreted by the application of mathematical methods." Aristotle (384-322 BC) on the other hand, had a perfect interpretation of the meaning of the circle! The many other scientists, physicists and theorists of all times with their useful contributions, encouraged and helped us to proceed further more, with our knowledge on physics and theories to solve the mysteries of our cosmology. I think that we the humans should be convinced that all these events of our misleading world of nature are as they should be.

We may call all these happenings mysterious and/or paradoxical or even that God wanted things to be made that way. But this is a reality of a world not real, made out by necessity, by an intellectual like nature. Following the path of my cosmic theory, of how the universe has been made, I will go ahead with some details of how the first atoms have been produced, and how the building of the universe had started. I will go back to the production of the particles made out by leptons. These were the trios, quarks, protons, neutrons, etc., and they were the first made almost automatically to create and develop a universe as it is observed today. It was the first and the most important stage of our universe creation using the same heat power that was used during the big bang. I will explain the building up and development of the smallest and plainest microcosmic planetary systems of the hydrogen atom isotope of protium and so forth. Starting up this summary of the universe creation, I am not to repeat everything that has been said previously again about how the leptons were created. You may find all the details in chapter 6. I am going to say though a few more interesting things about the enigmatic quarks and the very important and omnipotent proton and neutron that made the masses of all the stars and galaxies in our universe. The quarks were the pro-runners of the protons and the neutrons. Those quarks at some cosmic time during the critical moments of their construction in the first stage of our universe, had reached their own size of measurement, not growing any more. The neutral quark as I have explained it before in chapter 6 attracts a negatively charged trio to produce a negatively now charged quark heavier by three more leptons: (1e⁻) + (2v) down quark. The third quark is the positively charged one (the up quark) that was ready before it had reached the age of the neutral quark. It was lighter by one trio of (1e⁻) + (2v) negatively charged. It had 609 leptons altogether, while the neutral had 612, and the down quark had 615 leptons consecutively so that altogether there would be 1836 leptons that are contained in the proton but they all, being different from each one to the next by 3 leptons, and so they could develop a fractional charge of 1/3 among themselves!

One may wonder why all of the quarks didn't follow the same pattern or the same procedure so all of them could be produced in one size instead; for instance all of them to be of a neutral size of 612 leptons or 615 or 609 leptons each. How did it happen that they were all connected at the right time and with the right curious number of leptons in each one of them? That is something that makes you wonder. I will try to explain all of this in more detail as much as I can. You should now remember that:

a) The pre-existing mass is consisted of a certain ratio of heat and cold power, and of course it must be consisted of a total proportional number of leptons that will fit the cosmic equation number 1 (two in one), representing the whole universe; which universe is the same total pre-existing mass (see chapter 6). I will also remind you here that 9 of these leptons will be the minimum number of leptons to reflect a microscopic universe with the same proportions of heat and cold powers as that existing in the whole universe. These leptons and the particles of trios and the heavier particles of trios up to the point where the neutral quark is produced can only live in very high absolute temperatures, of billions of degrees Kelvin. The magnitude of this very high temperature is only present in the first picoseconds or nanoseconds in a nuclear furnace of the big bang of an embryo universe. While the very low temperatures could only be found towards the very end of our universe.

b) The neutral quark will be the main issue of producing the other two quarks and the three together to produce the neutral proton first, and the unstable neutron after. These protons and neutrons whether you believe it or not, will construct all the masses of stars and galaxies of a misleading world of nature, and the more significant thing yet is that they are going to live as long as the universe will.

You probably now understand that the neutral quark was not made at random it was made by necessity. The leptons and the particles of trios were not allowed to live by themselves beyond a certain degree of a lower intensity in absolute temperature at that cosmic time. At the same time there was a proportional limit of numbers of leptons that were going to fit a neutral quark and of course to fit also the equation of:

1/3 (e⁻ + e⁺) + 2/3 (v) = 1, which is the number 1 of the (two in one) cosmic equation. In my opinion these two things were the main ones that would have made this work. The last wondering thing is how and when a neutral quark could find the other two quarks made out of a different number of leptons each so they can link together and become a nucleon of proton? Well in my opinion, when these neutral quarks were produced in large numbers there is a possibility that they were repelling each other for sometime until a new negatively charged quark would make its appearance, and do you know how this new negatively charged quark was made? It was made simply by an added negatively charged trio connected to that neutral quark, so that to change it in to a down quark. In the book "Building the Universe" by Christine Sutton on page 200, she says "a neutron is disintegrated in about 15 minutes to be replaced by a proton, an electron and an antineutrino. This is the so called beta decay of neutron." I say that in this case the neutron becomes free of a negatively charged trio of a $(e^- + 2v)$ as an unstable particle and becomes a proton. But let us not forget that this neutron is made out of three quarks, and in my opinion the down quark could disintegrate in 5 minutes, just 1/3 of the time needed for one neutron to disintegrate in 15 minutes. Also known is that this quark is a third of the mass of a neutron.

All these said above are fundamental issues and important laws of nature and I don't think that they could ever be altered as they are all connected with the ultimate being of our existence. As I said above, the down quark in my opinion could disintegrate in 5 minutes and become a neutral quark as it was before. Also this is the first light part of our universe creation. As soon as the quarks were completed and the famous proton was produced, the second part and the heavier construction of the universe would start. Since the proton was generated, a whole universe of stars and galaxies were produced. The universe has been getting a new form. This proton now has been figured out and analyzed. The entire world of secrecy was hidden behind that proton. This proton could never be alone; it has to have a company of an electron that will be spinning around it. Coming back to the first moments of creation, there was still a high intensity of temperature in billions of degrees in that embryo universe, the produced neutral protons still had the chance of grabbing female trios, that were still existing in abundance. The protons could not afford to live by themselves altogether. They could only repel each other until they were probably annihilated. So they worked out a way to get children and stabilize themselves by developing microcosmic planetary systems. How did they do it? Well, actually it was very simple. In every region of the still young and hot universe the protons were produced in abundance by the quarks. They were in a galactic space where the mass was in an intimate contact. It was indeed a filled up space of protons (a protonic space as it is named by other theorists). But there was still a female trio $(1e^- + 2v)$ for every proton in the universe to be connected with and fulfill the cosmic equation:

$1/3 (e^- + e^+) + 2/3 (v) = 1$. Then an unstable neutron was produced (breaking down in 15 minutes). However, I think that this is how all these things were happening in that critical cosmic time! And in this case most or even the huge numbers of protons were easily transformed to neutrons. It was like they knew what they were doing at that cosmic time. The protons having reached the age of making atom families in the space, they started to get pregnant by connecting themselves with the immeasurable clouds of the female trios in the space and transformed to neutrons. Every proton of 1836 leptons, which represents the number 1 of the (two in one) cosmic equation of:

$(306e^- + 306e^+) + 1224v = 1836$, needs only $(1e^- +2v)$ to totally change to the number 2 of the (two in one) same equation, and get the neutron construction of:

$(307e^- + 306e^+) + 1226v = 1839$.

You might imagine what has happened now. The whole space of the young universe would change very rapidly to a "NEUTRONIC SPACE". As it has been named again by other theorists of how creation has started. It was a dramatic and fast change that has been accomplished in this critical cosmic time. Even at the beginning of this transformation, the rapidly changed protons to neutrons would not have any chance to be attracted by other protons nearby which were of a pure neutrality, although these neutrons were slightly negatively charged after their transformation from protons to neutrons, just an instance before. And you may wonder what was the reason not to do so? Well, the protons would prefer to attract the more negatively charged particles of the trios in space, which were in a form of $(1e^- + 2v)$ and they were even much lighter than the neutrons. While these interactions were taking place in billions of Kelvin degrees in temperature, the time was passing by, and this made it surpass the 15minutes at which a neutron is forced to break down! This was a very serious event for our cosmology and for our existence in this misleading universe. **AT THIS MOMENT THE NEUTRONS WILL GIVE BIRTH TO CHILDREN**, and with a smart trick, they will be transformed to families of protium atoms of microscopic planetary systems, and stabilize themselves and the universe once and for all. This event will set the turning point of the foundation and the construction of our universe by making the families of the microscopic planetary systems of the world, and later the larger massive world of stars and galaxies; all being in the form of planetary systems in general. The universe now will start to change its spectrum. The first atomic families will make their appearance. The neutrons were overloaded just a few minutes ago, and by nature they looked like they were pregnant particles until a little while after, when they got rid of their load. What an intractable universe it is indeed!

THE FOUNDATION OF A HEAVIER UNIVERSE

As soon as the neutrons started to break down, all the regions of space had been changed and filled up with hydrogen atoms; the first planetary systems in the world and plainest atoms of protium isotopes that were ever created. This whole thing looked very easy, but at the same time it was a very smart trick. As the breaking down of the neutrons had started simultaneously they were giving birth to the children of one electron (e⁻) and two neutrinos (v), and at the same time the neutrons were changing their status and from particles of nucleons or hadrons as they are called by the particle physicists, they became particles of nuclei of the protium atoms! (The same protons changed in to neutrons and back again to protons). They became the mothers now so they can take care of their born children, one electron and two neutrinos, for every mother proton. These children looked like they were supposed to stay with their mother, but that was not the case; the mother proton could not attract and get hold of all three children around her. The mother nucleus of the neutral proton now, succeeded to attract only the electron (e⁻), the weak daughter but not the more powerful neutrinos (v). The attraction of this electron, was accomplished by an electromotive force in a microcosmic distance between the proton and the electron, measured in some billionth of a billionth of a meter which would create an astronomical magnitude of a force of attraction, and this because of the extremely small distance separating them, as I will write about this later.

That was the electron (e⁻) with a fractional charge of 1/3 minus, which was an easy prey for the mother to attract, while the neutrinos much more powerful with a neutral fractional charge (the same as the mother), were repelled and slipped away very easily. That is why neutrinos were in abundance in the earliest cosmic times, traveling allover the space, with almost the speed of light and not easily influenced by any force of attraction. The volume of space in the earliest cosmic times was much smaller than it is today, and they could be easily located while at present neutrinos are spread out to a very larger space in the same universe and they are rarely located. The neutrinos all over in our universe had an objective. It was a clear task that was to be fulfilled by them. I think it was like a duty for them to undertake the task of controlling their sisters and brothers, (electrons and positrons) and to restrain them from fighting each other. Because if they remained free they could come face to face confronting each other and this could result to their annihilation. In order to avoid this bad habit, by necessity the neutrinos used a very smart trick. They were always going in two at a time, to grab an electron (sister), or a positron (brother), and they were getting a strong hold on them, making the whole universe remain in harmony, and get rid of these problems. Another significant thing that could have happened at that cosmic time, is the possibility that those developed atoms of protium isotopes were being fused with an unprecedented number of neutrons, before the latter were broken down. Thus they could form the deuterium atoms of the hydrogen isotopes. At this point another important thing should be taken into account in this paradoxical, multilateral and multiform universe. The neutron of course has changed and it has been transformed to a protium atom, a family of a local thermal equilibrium, but this microcosmic family being the first to be produced in our universe and the smallest and plainest, is far from being perfect indeed!

To create a universe with true perfect conditions in it could never be possible. The nearest point to succeed such a condition is when we have a space filled up with neutron particles that are not broken down. This could not be possible. Even at the very "END" of our universe in its absolute zero degree where it is transformed to a pre-existing mass, and an instant after this, another big bang would occur. The light formation of the creation was up to where the quarks, the protons and the neutrons were all done. ALL OF THEM WERE MADE DURING A CRITICAL COSMIC TIME I believe. The heavier part of our universe started with the development of atoms of the hydrogen isotopes, which are protium, deuterium, and tritium. Also there are other heavier elements. The formation of the atoms was the first planetary families to stabilize the universe and my point of view is that they will never be perfect. The nuclei of all protium atoms will always be missing the two disappeared neutrinos. The electron will only be turning around, but it will never succeed to link up with the mother, as long as the hot and cold temperature is above zero Celsius degrees and the universe is in action. That is why the universe is alive for billions of years. This first family of the protium atom though, might be transformed to any other isotopes of heavier atoms by attracting other neutrons and get fused with the nucleus of this protium atom in extremely high degrees of temperature as the particle physicists say. In this case, the neutrons will be able to live together with the protons and strengthen the cores of the nuclei in the atoms and make them heavier. These protium atoms could become isotopes of deuterium or tritium with the same number of protons in their nuclei (just one), and an extra one or two neutrons fused with them, to change their form and make them heavier and more negatively charged, but this change could not produce a nucleus to be looked upon like the same neutron as it was before. In fact, generally all the nuclei of the atoms in our universe are made of protons and neutrons. If you make a good observation of all the atoms in the world, you will see that all these atoms first were nucleons of protons, before they changed to neutrons, and then to nuclei of hydrogen isotopes and other heavier atoms. In the end it could be postulated that no neutrinos or nucleons of protons and neutrons could live alone in our universe. Both neutrinos or protons being alone would fly apart, and the neutrons alone would break down in 15 minutes.

I think that these protium atoms are making up the highest percentage being present in the whole universe. The lighter atoms on the periodic table of elements are containing equal number of protons and neutrons in their nuclei, as the particle physicists have discovered (see the periodic table on pages 204-206 in the Encyclopedia Americana, 1997 edition) while in the heavier atoms there are more neutrons than protons in their nuclei. There are about three neutrons for every two protons. At the same time we should bear in mind that more neutrons than protons in the nuclei of these atoms, will change their status to a more negatively charged particle; but just a little, never to reach the previous state of a genuine neutron! Also these nuclei are becoming more massive and reinforced. I have explained all about positive and negative charge of the nuclei of the atoms in chapter 7, and this will give you more ideas of how much more negatively charged these nuclei could become when adding neutrons, or neutrons and protons together.

All these things may be confusing a little but this is what is going on in this universe, and we cannot change it. I may conclude then that nothing is equal and nothing is real in this cosmos. Only cold and hot temperature, and these two are unequal. That is why hot and cold temperature would never make a solidarity. That is why the one electron and the two neutrinos could never go back to the nucleus of the protium atom in space and at any cosmic time, except when temperature is below zero Celsius and there is an inactive universe, where the child of the electron is linked with mother only for the purpose to absorb temperature and overcome the thermodynamic power that had expanded the space of the universe before. Ending up this chapter, of the lighter creation and formation, and the heavier construction of our universe, I am convinced that a good understanding of the microcosmic world has been given at large, and a good comprehension of how the universe has been created and constructed up to this point. Next, in the "SECOND PART OF THE COSMIC PHYSICS", I will go on further relating to many other detailed events, and all the factors concerning a whole universe pulsating in the eons of time. These are the current and the voltage, the electric charge of the electron, and how all of these are functioning, as well as the heat power of the stars, etc. Also, I will mention the rest mass energy of an electron or a positron and the total release of energy in the whole universe. The force of attraction, and the equation for this force I believe will constitute the one equation of the "GRAND UNIFIED THEORIES". It is one of the hottest problems concerning the whole universe, where the four forces are reduced in one only force.

THE SECOND PART OF
THE COSMIC THEORY
OF THERMODYNAMICS

CHAPTER 13

NEW PRINCIPLES AND LAWS OF COSMIC PHYSICS-THE ELECTRO-MECHANICS OF THE CIRCULAR MOTIONS IN THE SPACE OF THE UNIVERSE

FROM THE BEGINNING of chapter 1 to chapter 12 in this book, I have extensively dealt with temperature and its thermo/cold-dynamic powers, the masses and their potential energies, the deep meaning of the absolute zero degree of temperature and the infinity, the epoch prior and after the **BIG BANG** of a closed universe, the creation of leptons out of their pre-existing mass (the mother of nature), and their detailed constructive appearance and intrinsic energy of heat and cold power and their fractional charges, ratios and proportions to the universe. How the quarks, protons and neutrons were developed out of these leptons, and later how the atoms, stars and galaxies were developed and grown up. I have also talked about the trinity and the doctrine of the metaphysics, how much cold and heat there is in the universe, volume, mean density and mean temperature. I have also described a summary of the universe with many details about the **BIG BANG** and Genesis. This was the **FIRST PART** of my cosmic theory. Now I will introduce the **SECOND PART OF COSMIC PHYSICS**. It will be an extensive chapter. I will explain how the universe is observed, studied and calculated.

The circular electro-mechanics of the masses in space in the universe is the first and the last thing of our **COSMOLOGY** to be studied thoroughly, in order that all the arising problems of the cosmic physics could be solved by the science of the mathematics. I think I have recently discovered that the whole universe including all energetic heavy and light masses of stars, galaxies, particles and even the light itself is living in circles which are tending to follow elliptical and/or spiral directions. The arising hot problems are many to be studied, especially the problem of how to find out the equation of the grand unified theories (G.U.T.) that so much has been discussed about this. I think that the solution is exactly as the scientists put it, of how the four forces in the universe are working. Then I would believe too that most of the problems should be resolved. Much I think is based on the classic electrical laws, especially on the equation of (I = E/R).

We the humans should be wondering how could it be that a whole microcosmic system of leptons, quarks, protons and neutrons inside the atoms, are working in harmony for billions of years releasing immeasurable amounts of energy internally and participating with the macrocosmic huge planetary systems externally by radiating, expanding or contracting the universe. Where does the power come from for the tremendous gravitational forces, making stars and galaxies orbiting one after the other and in the mean time cycling a spherical and finite universe? (See page 84 in the book "The Big Bang" by Joseph Silk, what he says about spherical surfaces). Physicists say the laws of the nature are the same for all uniformly moving systems. Which are those same laws in detail? Who is going to answer about the volts, the amperes and the EMF's in space? Who is going to answer about the forces of attraction and all the kinetic forces in space? Who is going to answer all of these and many other problems which all of us are witnessing everyday in our universe? The time has come for someone to find out about all these hot issues and problems for HUMANITY.

I made up my choice to start from the atoms, to explore them as much as I can and see how they are functioning in the depth of their infinitesimal confinement comparing them with our solar system of the external world as much again as I can, in order to reach my objective of how the whole universe is making out. I read books, encyclopedias, theories of the big bang and creation, news of scientific information, astronomers' magazines, etc. Many a times I was not satisfied about big problems and hot issues of our existence. I was trying to do it in my way using my human wisdom to figure out the big problems. Almost since 1955 I was imagining how I could answer questions about our cosmogony and cosmology. The motive that has impressed me to a significant degree to start my studies from the atoms was the electric charge of the electron. What sets the electron's charge at $1.6 * 10^{-19}$ coulombs? That is what Christine Sutton, editor of the book "Building The Universe" is asking on page 125. And why is its mass 0.51 MeVs and not 5 MeVs? These issues of cosmological importance have given me a hard time thinking about this and the reasons that made all the electrons having an electric charge of that much low measured in billionth of a billionth. What is the meaning of that electric charge? Why this electric charge was found to be $1.6 * 10^{-19}$? I was thinking about the kinetic energy of the electron, the thermo/cold-dynamic powers and their relation as well as other circular motions and their electro-mechanics in the whole space of the universe. But let me again examine first that same strange equation of (I = E/R) of the classic electrics. Is there a similar situation of producing current with the generator machines as that with the speeding electrons around their nuclei? I thought I knew current is produced by moved electrons in a conductor at high speed, and the number of the moving electrons running through in a second will determine the rate of power. The resistance (R) and the potential difference (E), or the voltage will regulate the amount of current flowing through the conductor and doing work for a certain time (t) in seconds. Also current may be produced in chemical batteries and other apparatus of microchips, etc. So far so good! All this is contained in classic physics. Not the beyond standard physics that I think I have discovered myself! The potential difference (E) or the voltage will regulate the amount of current flowing through the conductor and doing work for a certain time (t) in seconds. Resistors are made plenty and depend on the material cross section and length of the conductor. Is there now any difference in space though? What current is in space? How does it move? What current is it? How fast is it running? Is it the same produced current in the electric machines as it is being generated in space? Is there anything else that could make current too? What do we mean when we are talking about current and voltage in space? How are they generated? Where does the current come from? Where is it going? What is the direction of the current and the voltage in space? How are they measured? In classic electrics there are the following factors:

VOLTAGE

According to the scientific interpretations, the volt is a unit of electric measurement in the metric system as the international system of units. Its symbol is "V". One volt is the potential difference between two points, if one joule of work

(One joule = .7374 ft lbs.) is done in moving a charge of one coulomb between the points. A volt can also be defined as the difference in electric potential between two points of wire that is carrying one ampere of current, and producing one watt of power. Potential differences could also be called voltages, and they are related to the energy of the electrical forces that push charges through a conductor. One volt of potential difference across a resistance of one ohm produces a current of one ampere.

THE CURRENT IN AMPERES

The scientific interpretation of the ampere is that it is a unit used to measure the rate of flow of an electric current; it is one of the seven base units in the metric system and its symbol is "A". They say that the ampere is used to measure electricity much as the unit gallons per minute is used to measure water. The electric current flows at the rate of one ampere, when one coulomb flows past a section of an electric current in one second. One coulomb electric charge again is equal to 6.28^{18} electrons as it is established in classic physics. Physicists define amperes, in terms of the magnetic force measured in newtons.

RESISTANCE AND THE POWER IN WATTS

The resistance is measured in ohms and it is depended on the distance the current will travel, the cross section of the wire of the conductor and the material that this is made of. If the potential difference between two points is one volt, the resistance one ohm and the current equals one ampere for one second, the power of one watt is produced.

THE COULOMBS

The coulomb is interpreted by the electro-physicists as a unit of electric charge defined as the amount of electric charge that crosses a surface in one second when a steady current of one ampere is flowing across the surface. A charge of one coulomb is equal to $6.28 * 10^{18}$ electrons. For all of this mentioned above about the factors of classic electrics I only say this: the current in amperes, the voltage in volts, the resistance in ohms, the electric charge in coulombs and the watts are the same in space as that in the classic physics! But all the factors above are defined in a different way and value in space I believe. This is because the classical theory in physics has been changed by myself to another theory of physics, which is beyond standard model.

CHAPTER 14

THE CIRCULAR ELECTRO-MECHANICS IN SPACE

HOW IS THE thermodynamic power of a star or a galaxy calculated? How is the current and voltage generated in any planetary system? What is the meaning and the value of the electric charge and all other factors in space? Here I must deal with circular electro-mechanics as they are presented to us by the two existing planetary systems in space. They are the microcosmic and the macrocosmic ones (see chapter 10). After a few years of writing and erasing texts for my cosmic theory, I have discovered in my opinion the true meaning and values for the current in amperes, the volts, the resistance, the electric charge of the electrons, the mass converted to energy of electro-volts, and other important issues of principles and laws in the circular electro-mechanics of the cosmic physics in space. I feel I am very satisfied in making a significant progress in these fundamental issues of cosmic physics for the humanity.

In space there are positively charged thermodynamic heat power sources of mass, (hot stars for instance) that could release tremendous thermal kinetic energies marked as (Ke). If I symbolize the power of a star or a galaxy with the letter (W), the heat power with the letters (HP), and the speed of light with (c), the power of a star or any body of mass in space is equal to: $W = HP * c^2$. The heat power (HP) of these stars is equal to the tM * mT (total mass * mean temperature). See chapter 5. The equation above for the power of a star could read: $W = (tM) * (mT) * (c)^2$. This thermodynamic kinetic energy released could be changed to a current in amperes (A) of classic physics, if and only the star was connected with a planet, and form a planetary system in space, like our solar system. This kinetic energy of the sun for example, could change to a positive current (I), and it would be losing its strength to the square meters of the distance, while it was traveling in space until the time it had touched the negatively charged object of mass to form a thermal local balance or thermal local equilibrium. That is how current is produced in space and with such strength that this thermodynamic (Ke) of the speed of light in the square, could make the planet earth be pulled around the sun (this force is called FORCE OF ATTRACTION). This powerful (Ke) of the sun could cause a sequence of events with any of its planets, or other objects of mass shaping a closed electrical circuit. I will discuss later more about electro-mechanics, which are the values and the meaning of all electric factors. We should only remember ($I = E/R$) of classic physics. I will come back to this equation many times. All of these important issues I am going to talk about should be thoroughly understood. I would like to talk extensively about the sun-earth-moon and nucleus-electron, their electric charges (el-ch), their revolutions per second (Revs/"), the time for one revolution, the forces of attraction (FATT), and generally all the other issues concerning them as a whole. And of course how did I discover them all?

But indeed how did I start, in order to go on the right path and move ahead with this theory of cosmic physics? How did I work it out from the beginning to complete a very difficult task? How hard was it for me to concentrate and investigate important physical events and phenomena in our universe, which were linked to our cosmology? How did I determine the number of leptons for instance in an atom and the amount of their electric charge? First, I came to the conclusion that this electric charge belonged to all kinetic objects or particles of mass in space in the globular universe, and not only to the leptons! The Mother Nature gave birth to the kids of leptons during the first picoseconds of the big bang in our earliest cosmic times. Since then, these leptons worked too hard to create their first families of protium atoms, and later the heavier ones in the universe. They have built up their own microcosmic planetary systems of local thermal equilibrium, and they started to work there like families, for billions of years releasing an immeasurable amount of energy since then. I looked at these atoms to find out how they were working there, and what amount of energy they might release in one second in order to proceed further on, and find out what energy is released by the heavier objects of mass per second in space, which after all every one of them is consisted of a multiple number of leptons as I have explained this in chapter 6. I thought it would have been much easier to start with the atoms, and try to find out what is the meaning and the value of the electric charge of the electrons in their own microscopic world, and I think I was right! I knew the chances were I could get more positive results starting with the plainest atom of the protium isotope. On the other hand I was confident I was on the right path in my cosmic physics of the "Thermodynamics" which are related to the "Cold-dynamics". Reading and studying the encyclopedias, and searching the scientific news on the computer, I found many interesting news. One of these is the G.U.T. (Grand Unified Theories) attributed to Dimitri Nanopoulos, a theorist of Harvard University, who published a paper in 1978. For myself, there were two other interesting things that I paid attention to. The first was about the electric charge of the electron, and the second was the diameter of the atoms. This electric charge seemed to me like a strange physical phenomenon. I started to think about it, and I was very anxious to figure out a logical solution about this mysterious problem. More pressing on me at the time were the words of Christine Sutton, editor of the book " Building the Universe". On page 129 she said " I don't think the tau will help us to answer questions such as: how can we calculate from basic principles the magnitude of the electric charge (1.6 * 10^{-19} coulombs) possessed by all the charged leptons? This question, like the question of how to calculate the speed of light from basic principles, is so intractable that it must be left for future generations of hopefully brighter physicists."

I started with the diameter of the protium atom. I am not a physicist by profession, but I knew the atom is incredibly tiny, (page 633 from the Encyclopedia Americana, 1997 edition). It says atomic diameter is: 2 * 10^{-8} cm; fifty million

(1/2 * 10^{-8}) atoms are ranged side by side in one linear cm of solid matter! A cubic centimeter of any solid will therefore contain about 10^{23} (one hundred thousand billion billion) atoms. I started to think about these millions of atoms, and I was only trying to find out by imagining how an electron was spinning around its nucleus, and comparing it to the earth circling around the sun. The most significant thing to investigate was the fundamental issue of the electric charge of the electron. I paid my attention to the amount of this electric charge, as set by the particle physicists equal to 1.6 * 10^{-19} and to the meaning of the amount of one coulomb (as a unit of electric charge), which is supposed to be equal to 6.28 * 10^{18} electrons that cross a surface in one second, when a current of one ampere is flowing across that surface, according to the particle physicists again. Taking into account all these that I mentioned above, I came to the conclusion and made up my mind that the electrons orbiting around their nuclei with almost the speed of light as well as their infinitesimal mass and distance between nuclei and electrons, play a very important role concerning their electrical charge. The diameter of about 2 * 10^{-8} cm should be measured to billionth of a billionth of a meter, and the electric charge of the electrons should also be measured to billionth of a billionth in meters too. The electric charge estimated by the particle physicists of 1.6 * 10^{-19} coulombs; I thought it was close to what it needed to be! Also I should not forget, that this electric charge is developed by an electron moving around its nucleus in a circular or elliptical orbit and the constant (π) and the (radians) should be taken into account. This situation took me a long time to think and find out what were the numbers I needed, using the mathematical laws of the circular electro-mechanics I thought I was doing good progress. In the end, I satisfied myself, I came out with this: the electric charge of the electron in an atom, is a negative resisting power (in ohms) of the mass of the (e^-), expressed in an elliptical kinetic energy around its nucleus. The nucleus contains the positive power, which will attract the electron, (the same I believe like the sun is doing towards the earth). The amount of this electric charge of the electron then in ohms should be equal to the mean radius between the nucleus and the electron in meters, multiplied by one radian. For the electron in the protium atom, it should be equal to .00000000000000001 * $1/2\pi$ = 1.59154943092 * 10^{-19}. This was amazing. In my opinion this is the accurate electric charge of the (e-), when the temperature around the atom is 1^0 Celsius. I had made up my mind that in 1^0 Celsius the radius between the (e-) and the nucleus should be 1 * 10^{-18} because I did find out that everything was working right. (See later why the temperature should be that). This is very close to what has been estimated by the particle physicists. The mean radius between the nucleus and the electron is changed with any changing temperature around. I will come back to this later. The nucleus has a positive power pulling the electron around. This is called force of attraction (FATT), and this force of attraction is developed because of a difference in heat powers between the nucleus and the electron or between the sun and the earth-moon. But how much distance in meters will the electron or the earth-moon travel on their elliptical orbit before they stop resisting and are pulled around? You might not be surprised by what the answer is; it is just one meter! No more and no less. No matter if it is an (e-), or the earth-moon or any other body of mass as a one unit, turning around another body of mass, including our solar system orbiting around the Milky Way. It is again a distance of one meter. But pay attention to this: for one meter distance the

(e-), I believe will complete

$1/(2\pi * .00000000000000001)$ = 1.59154943092 * 10^{17} revolutions, and these revolutions would be completed in one second when the temperature around the atom is 1^0 Celsius while the earth-moon with an approximately mean radius or mean distance (mDist) of 149538928746 meters from the center of the sun to the center of the earth will have $1/(2\pi * 149538928746)$ = 1.06430442178 * 10^{-12} revolutions completed in one meter travelled distance, and if the speed of the earth-moon is approximately 29773 meters per second it will take 1/29773 = 3.3587478588 * 10^{-5} seconds to travel this distance of one meter. (Distances between the sun and the planets taken from page 11, 101, 230, 355, 718, 799, 800 from the Encyclopedia Americana).

HOW THE PLANETARY SYSTEMS ARE FUNCTIONING IN SPACE AND THEY ARE GENERATING THE CURRENT IN AMPERES

The current is only produced inside the circuit of a planetary system in space, between two objects or particles of mass, like it is generated in classic electric physics between two points in a conductor. As an example, let me take the same planetary systems between the sun and the earth-moon, or the electron and nucleus of an atom. (The earth and moon taken as one body of mass). One of them with the higher heat power source (HP) will be the positive charged object, in this case it will be the sun; and it will generate the positively charged current (I) in amperes (A), while the other with the lower heat power source (hp), will be the negatively charged object (the earth-moon in this case), and it will produce the negatively charged current (I). That negative current will be produced by the kinetic energy of the earth-moon's mass, turning around the sun or the electron's mass, spinning around its nucleus. The potential difference (HP-hp) between the sun and the earth-moon or the nucleus and the electron will cause a sequence of reactions to both of them that will be (EMF), (FATT), (el-ch), etc. The electromotive force (EMF) will develop a force of attraction (FATT) which will generate the positively charged current (I), starting from the sun or the nucleus, and going through to the (e⁻) or to the earth-moon. This positively charged current would have the power to pull the earth-moon and/or the (e⁻) in orbit around the sun or the nucleus. It will be an elliptical kinetic energy (Ke), by the earth-moon or the (e⁻) producing a negatively charged current in amperes which will be returned back to the sun or to the nucleus, balancing the agitated space and creating a thermal local equilibrium.

It would be a continuous flow of current in amperes between them, and lasting for billions of years, but this will be done not before there would be a negative resisting

(el-ch) initiated by the mass of the earth-moon and/or electron. These electric charges will be equal to (mRd) * $1/2\pi$, (mean radius distance multiplied by $1/2\pi$) as it has been mentioned before. They are resisting powers in ohms against the (FATT), lasting one meter distance travelled. Later I am going to explain how I have discovered that this traveling (Dist) is exactly one meter, or ($1/2\pi R$) revolutions. The time in seconds to cover this traveling (Dist) of one meter is very much different in any other pulled around object or particle. But how is this electric charge developed and why? The (el-ch) is necessitated by the force of attraction and the inertia of the mass of the pulled object or particle. This is true because the mass of the (e⁻) and that of the earth-moon is destitute of the power of moving itself. It is not having or possessing wanting of moving. The inertia of their mass causes a resisting power generated by themselves in ohms opposing the (FATT) for a steady travelled distance of one meter, and for a short period of time in seconds. The (FATT) is the result of a potential difference between the two or more objects in either external or internal planetary system; external planetary systems is using great distances starting from one meter and up, while internal systems working in decimal numbers from unit one meter and down. For instance the earth has a mean radius from the pole of the sun to the pole of the earth about 149538928746 meters and for 1 revolution it is:

2π * (mRd) = 939580799950 meters, while the electron has a $1 * 10^{-18}$ meters mean radius distance from its nucleus in 1^0 Celsius and 2π * (mRd) = 6.28318530718 * 10^{-18} meters for 1 revolution. The electric charge (el-ch) of the (e⁻) is $1/2\pi$ * R (as mentioned before) or $1/2\pi$ * $(1 * 10^{-18})$ = 1.59154943092 * 10^{-19} meters of kinetic energy in ohms. If I divide the (Dist) of one revolution of the (e⁻) by its (el-ch) I get

39.4784176044 electric charges in only one revolution of the (e⁻). This is the same as the $(2\pi)^2$.

For the earth-moon again the electric charges for one revolution are:

$2\pi R / (1 * R / 2\pi) = 2\pi * 2\pi = 39.4784176044$. The traveling distance now for 1 meter converted in revolutions for both the electron and the earth-moon will be: $1 / (2\pi * 10^{-18}) = 1.59154943092 * 10^{17}$ revolutions for the (e⁻) and $1 / 2\pi R = 1.06430442178 * 10^{-12}$ revolutions for the earth-moon. Another significant event in this theory is that the number of the total electric charges of the electron traveling 1 meter (Dist) in one second is:

$(2\pi)^2 * 1.59154943092 * 10^{17} = 2\pi * 10^{18}$ or $6.28318530718 * 10^{18}$, all of them done by one electron! But this is equal to one coulomb electric charge and I think this is what is needed for the (e⁻) to produce 1 ampere, and if it is done in one second the power of 1 watt is made; and I think this is exactly what is happening here. In the protium isotope of the hydrogen atom when the temperature in the neighborhood of the atom is an average of 1^0 above zero Celsius. How many electric charges could be completed by one electron in the planet of the earth during 1 revolution around the sun? And how many electrons there are contained in the total mass of the earth? You may find it after. I found out that the time in seconds for the earth-moon to complete one revolution around the sun is 31558150 seconds approximately, then the speed per second for the earth-moon should be about $2\pi R / 31558150 = 29773$ meters per second. Converted to revolutions, it would be $29773 / 939580799950 = 3.16875355495 * 10^{-8}$ revolutions per second. They are travelled meters / total travelled distance or traveling seconds / total travelled seconds = revolutions. Now here I will give you more explanations about the resisting power of the electric charge of the earth-moon or the (e⁻). This (el-ch) of the earth-moon and the (e⁻) as you may know up to now, is a (Ke) of $1 / 2\pi * R$ and both the (e⁻) or the earth-moon would have to travel 1 meter distance exactly on their track around the sun or the nucleus to develop this resistance. The (FATT) by the sun or by the nucleus in the protium atom, can also be expressed in amperes of the speed of light in the square. This is the result of the (HP-hp) $* c^2 / (mDist)^2$. (See chapters 15 and 16 for the FATT). The (HP-hp) $* c^2$ indicates power of current in amperes expressed as a forcible speed of light in the square; and depended on the (tM) and (mT) (total mass and mean absolute temperature) of the sun and the earth-moon, or of the nucleus (n) and the (e⁻), and their (mDist)² between them. In a closed electrical circuit the total resistance of the earth-moon for 1 revolution around the sun, is found by multiplying the resisting power of the earth in ohms for 1 meter, (which is the electric charge of the earth), by the total traveling (Dist) in meters of the earth in 1 revolution (which is the $2\pi R$). But this is the same resistance as that of the speed of light in the square of the sun. It is the (mDist) or the (mRd) in the square between the earth and the sun, equal to

149538928746^2 and expressed in ohms. This could be verified by the equation of: $(1 / 2\pi) * R * (2\pi R) = R^2$.

The $1 / (2\pi R)$ corresponds to 1 meter and the $2\pi R$ to the total traveling (Dist) of the earth-moon around the sun. This all is done because when the sun and the earth-moon are connected to a closed circuit, like it is in a classical electrical machine, the resistance in the line of the circuit is the same for the positive or negative current flown, except that here in space the total resistance per second of the speed of light in the square by the sun is counted in a different way than that being counted by the earth-moon. The faster moving speed of light in the square by the sun, traveling in almost straight lines towards the earth-moon takes less time in seconds to develop its resistance than the relatively slow moving earth-moon, which takes a longer time in seconds to complete the greater distance in meters of one revolution around the sun. The earth-moon could not develop its total resistance before it will complete a total traveling distance of one revolution around the sun. (See the next chapter). In each one complete revolution the earth-moon must produce a number of negatively charged amperes by its (Ke), which multiplied by its developed resistance would equal the supplied power of the sun.

The total power of the sun is $HP * c^2$. The resistance of the earth-moon around the sun for one revolution is $(1 / 2\pi) * R * (2\pi R) = R^2$. The (FATT) or the amperes per second imposed on the earth-moon by the sun is equal to: $(HP-hp) * c^2 / (mDist)^2$. So, if you multiply the (FATT) or the amperes per second by the resistance you will get the total power of the sun. See chapter 15.

The resisting power of the electron's electric charge for 1 revolution (when the temperature around is 1^0 Celsius) is calculated in the same way as it is for the earth-moon above. It is:

(el-ch) $* 2\pi R = (1 / 2\pi) * (mRd) * (2\pi R)$, or it is:

$(1 / 2\pi) * (1 * 10^{-18}) * (2\pi * 1 * 10^{-18}) = 1 * 10^{-36}$. This is the resistance for 1 revolution of the (e^-) around its nucleus. For $1.59154943092 * 10^{17}$ revolutions per second of the

(e^-), its resistance should be:

$1 * 10^{-36} * 1.59554943092 * 10^{17} = 1.59554943092 * 10^{-19}$. The revolutions per second of the earth-moon as it has been said, are $3.16875355495 * 10^{-8}$, if multiplied by the electric charges of $(2\pi)^2$ I get $1.25097376127 * 10^{-6}$ electric charges per second. One electron or one lepton on the earth independently turning around the sun would need $2\pi * 10^{18}$ electric charges being crossed in one second to produce one ampere. On the earth it should produce $(1.25097376127 * 10^{-6}) / (2\pi * 10^{18}) = 1.99098657784 * 10^{-25}$ amperes in one second. The planet Pluto with a traveling distance of approximately 248 years around the sun has a mean speed in meters per second about 4738.9 meters and (mDist) from the sun about $5.90251256437 * 10^{12}$ time to complete one revolution is approximately 7,825,989,200 seconds. The traveling (Dist) in meters is about 3.70865802199 * 10^{13} meters. The revolutions per second are:

4738.9 / 2πR = 1.27779373884 * 10^{-10}, multiplied by $(2\pi)^2$ gives us:

5.04452748342 * 10^{-9} electric charges per second, and divided by 2π * 10^{18} there will be produced 8.02861484549 * 10^{-28} amperes in one second. I found these distances in kilometers and the time in years from encyclopedias and have converted them into meters and seconds. If we compare the electrons spinning around their nuclei on the contrary they complete billions of revolutions measured in a millionth of a second. Here the total traveling (Dist) of one meter, and the total number of revolutions for one meter by the electrons are completed in one second exactly. The force of attraction, which is the speed of light in the square, in this case, and every other case, is transformed to a speed in revolutions per second by the (e⁻) or any planet, etc., which has an infinitesimal amount of mass equal to 9.1093897 * 10^{-31} kg (page 694 from the Encyclopedia Americana).

I recall in each one revolution the (e⁻) would cross its electric charge $(2\pi)^2$ times inside its atom. The (Ke) of the (e⁻) shown as amperes multiplied by a resistance would fill up the vacuum created because of a potential difference and balance the supplied positive power of the nucleus (HP-hp) * c^2 created between it and the (e⁻), and balance the power at any moment. The electron traveling a distance of one meter will have to make 1 / (2π * 10^{-18}) = 1.59154943092 * 10^{17} revolutions and then

$(2\pi)^2$ * 1.591549423092 * 10^{17} = 2π * 10^{18} electric charges should be crossed over in one minute time, producing 1 ampere * 1 ohm in one second (temperature around $1°$ Celsius). The number of amperes and the developed resistance per second by the earth-moon would be depended on the number of total leptons in their mass and the time in seconds to complete one revolution around the sun. The mass of the earth-moon is consisted of an inconceivable number of leptons speeding around the sun continuously, trying to bring back a balance in space caused by an irregularity because of a huge vacuum or voltage created from a potential difference between them. These leptons make the atoms and it is the same total mass of the earth and the moon in kilograms, and this mass is going to produce a certain amount of amperes turning around the sun.

Simultaneously the electrons, which are spinning around their nuclei, would produce extra amperes. This situation is made by necessity among the billions of stars and galaxies and other celestial bodies in space. The speeding earth-moon around the sun with their (Ke) will produce a negative current (I) in amperes and a resistance in ohms, which multiplied together at any moment or at the end of one revolution would balance the power of the sun supplied for the earth-moon only. It is only a partial power of the sun to be spent exclusively for this planetary system between the earth-moon and the sun. The total power of the sun has a capacity to attract all of its nine planets around it. The equation for the earth-moon orbiting the sun is: $(HP-hp) * c^2 = (I) * (mDist)^2$. The first part indicates the power of the sun in amperes imposed on the earth-moon. The second part shows the (Ke) of the earth-moon expressed in amperes multiplied by the resistance which is to equalize the power of the sun at any moment. This equation is similar to the electrical one of $E = I * R$ from classic physics. Here the (E) is represented in space by the (HP-hp), the current (I) is the created $(HP-hp) * c^2$, and the (R) by the square meters of the (mDist) between the sun and the earth-moon.

Any planetary system in space will be functioning in the same way. The supplied power in amperes per second by the sun or the nucleus, to the planet or to the electron, for any moment is equalized by the latter instantly using their own (Ke). I will give more details about this event later. The atoms with more protons and neutrons in their nuclei, will function the same way as the sun does. The neutrally charged mass of each one proton in the nucleus will have enough power to act and attract ONE ONLY negatively charged (e-). The mass of several NEUTRAL PROTONS in the nucleus will pull around equal number of negatively charged electrons. Here the nucleus represents the sun and the stars, while the electrons represent the planets around the sun and the stars. It wouldn't matter how far a spinning electron is from its nucleus.

Multiplying the useful work of amperes produced per second by the useless resistance developed with the mean radius measured in square meters, it will equalize the same thermodynamic power of one neutral proton. (Not the total neutral protons in the nucleus, because each electron will need another proton to be pulled around.) The same exactly performance is in the external solar planetary system, and other stars and galaxies in space. Except that here the planets, the stars and the galaxies while being pulled around in addition to the current produced per second by their total number of leptons in their mass, turning around other stars, there are more extra amperes produced in the atoms by the electrons which all of them are the same leptons that constitute the same mass of the planet. Pay attention to this now! The planet Pluto much further from the sun during one revolution will produce much more extra amperes in the atoms and less amperes with its mass than that of the earth-moon being much closer to the sun that will produce more amperes with its total mass turning around the sun in one revolution, and less amperes worked out by the total of its atoms inside the earth-moon. The FATT in amperes or watts per second, is equal to $(HP-hp) * c^2 / (mDist)^2$.

I would infer that in this situation there might be a relationship between the more amperes produced in the atoms by the planets farther from the sun, and the fewer amperes produced by the planets closer to the sun. Taking their total mass now as one bulk of leptons, the planets further from the sun will produce less energy than those that are closer to the sun which will produce more energy. That looks like either having a shorter or greater (mDist) from the sun, all of the planets could produce the same energy while orbiting the sun just for one revolution. Counting the total leptons, including those leptons, which are shaping the atoms in each one of the nine planets around the sun in our solar system, I found out the amperes produced by the nine planets equal to: $1.04595403804 * 10^{29}$, approximately when each one of them turning around the sun is completing one revolution. Their total amount of resistance in ohms at the same time, found to be about $5.50662171194 * 10^{25}$. If I multiply them two I get

$5.75967321556 * 10^{54}$. That should be the total power in watts per second produced by the sun.

I recall that this total power would be equal to $(tM)*(mT)*c^2$. (See Chapter 13). These numbers above would not necessarily be accurate. The (HP) of the sun could influence all other factors. It could increase or decrease resistance, amperes, etc. If I now want to calculate the amperes and the resistance produced by the earth-moon or the electron at any period of time in seconds during their traveling (Dist) around the sun or around the nucleus, this should be exactly the amount of power that has been used from the sun, or from the nucleus up to this particular moment. In other words, the power of the sun, destined for the earth-moon, will not be instantly absorbed by the earth-moon. Instead because of its heavy mass, the earth-moon will consume the power of the sun, little by little, and take some time before all of the sun's power, will be swallowed up by the earth-moon.

This time is approximately 31,558,150 seconds, and the earth-moon during this time will have completed one cycle around the sun. By contrast the electron, because of its infinitesimal amount of mass, and its micro-cosmic (mDist) from its nucleus, will only take one second to absorb the total power of the nucleus destined for it, and still in that short time of one second, it will be forced to complete $1.59154943092*10^{17}$ revolutions, or make it complete one revolution in $2\pi * 10^{-18}$ seconds around the nucleus when the temperature is 1^0 above zero Celsius. At that instant when one cycle of the earth or the total distance of the (e⁻) of one meter has been accomplished, their total kinetic energy of $(I) * (mDist)^2$ will equalize the (HP-hp) * c^2 of the sun and the nucleus. The first part of the equation shows amperes * resistance by the earth or the electron and the second part, shows the power of the sun or the nucleus. Figure No.14 on the following page shows the (e⁻) of the protium atom of the hydrogen isotope being activated in two different temperatures. That is 1^0 above zero Celsius and 4^0 above zero Celsius.

Fig No. 14

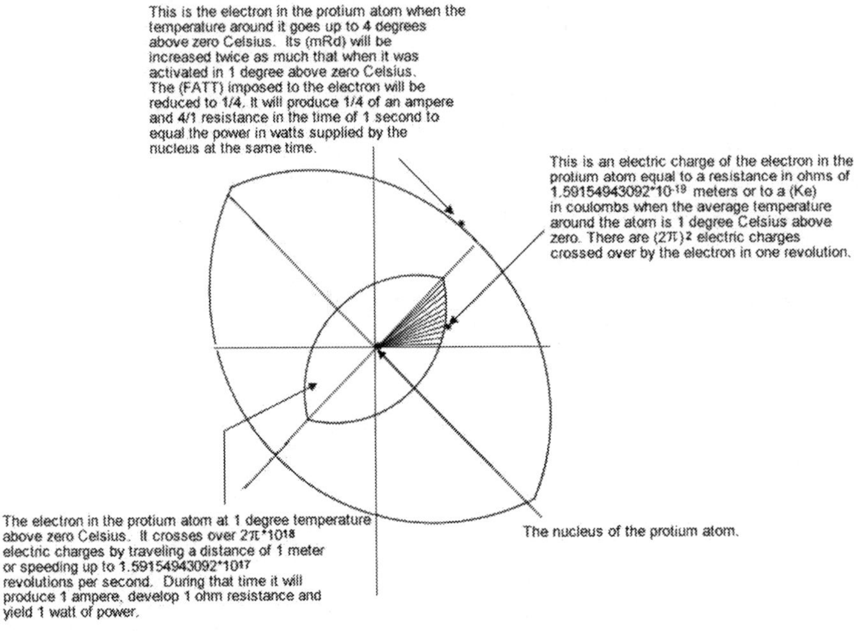

This is the electron in the protium atom when the temperature around it goes up to 4 degrees above zero Celsius. Its (mRd) will be increased twice as much that when it was activated in 1 degree above zero Celsius. The (FATT) imposed to the electron will be reduced to 1/4. It will produce 1/4 of an ampere and 4/1 resistance in the time of 1 second to equal the power in watts supplied by the nucleus at the same time.

This is an electric charge of the electron in the protium atom equal to a resistance in ohms of $1.59154943092*10^{-19}$ meters or to a (Ke) in coulombs when the average temperature around the atom is 1 degree Celsius above zero. There are $(2\pi)^2$ electric charges crossed over by the electron in one revolution.

The electron in the protium atom at 1 degree temperature above zero Celsius. It crosses over $2\pi*10^{18}$ electric charges by traveling a distance of 1 meter or speeding up to $1.59154943092*10^{17}$ revolutions per second. During that time it will produce 1 ampere, develop 1 ohm resistance and yield 1 watt of power.

The nucleus of the protium atom.

CHAPTER 15

THE SUN AND THE EARTH-MOON IN THE SKIES

I WILL GO ahead exclusively talking about the sun and the earth-moon, with their dimensions, their distances in space, and all other factors concerning many of their actions in the circular electro-mechanics, so that you may deeply understand them, and see how they are functioning in their own planetary system. Let me start first with their dimensions, what is their amount of mass, temperature in degrees, heat powers, etc. Searching encyclopedias, I found out the diameter of the sun, to be equal of about 1,392,000,000 meters, and its mean density equal to 0.26. I could then find out its total mass, if multiplying the volume of the sun by its density. It should be $4\pi*R^3/3*.26=3.67189011568*10^{26}$ kg approximately. (This is about 330,000 times the earth's mass.) And if I take a mean temperature per lepton inside the mass of the sun equal to $7,000,000^0$ Celsius approximately, the heat power (HP) of the sun would be equal to its total mass, multiplied by its mean temperature. That would make it about: $2.57032308098*10^{33}$. Multiplying the heat power by the speed of light in the square, ($c^2=300,000,000^2$ meters) it gives approximately a total power (tW), equal to $2.31329077288*10^{50}$. For now I will accept that this is the sun's power, although I think that it might be much higher.

These numbers above showing amount of mass, mean temperature, etc., for the sun, and later for other celestial bodies, are not necessarily supposed to be the right ones. There could always be a difference in estimating their values. If the mean temperature is (mT) and the total mass in kilograms is (tM), then the heat power of the sun (HP) is equal to (mT)*(tM). This would give a powerful light emission, or radiation of temperature to the interstellar space. Cooler objects of mass like weaker stars and planets could be attracted. Before I go further I would like to refer shortly to the equation of the forces of attraction, (also see next chapter). For this equation I use the (HP) and the (hp), as the heat powers of the positively and negatively charged objects in space. The sun's and the earth's for example. If the speed of light in the square is (c^2), the mean distance between the poles of the sun and the earth-moon or between any other celestial bodies of mass, would be represented as (mDist), and the (I) would represent the current in amperes, then the force of attraction symbolized as (FATT), would be: FATT = $(HP-hp)*c^2/$ $(mDist)^2$. This force of attraction is the (I) in amperes indeed. Let me suppose that the sun and the earth-moon shape a planetary system. With an earth's diameter of about 12,756,320 meters, (pages 534, 535, 539 and 541 for the earth and pages 11 and 12 for the sun in the Encyclopedia Americana, 1997 edition).

I was also interested about all of the planets around the sun and their masses, diameters, mean distances, their densities and speed per second. Many years ago I studied our solar system and probably found different numbers from those mentioned above from recent encyclopedias. So I found the earth's diameter from older encyclopedias to be about 12,756,320 meters and set a density of (1); its mass would be equal to about $4\pi*R^3/3*1= 1.08686308434*10^{21}$ kg. The moon's mass on the other hand, is the earth's mass divided by 81, according to scientific estimations. It is: $1.08686308434*10^{21}/81=1.34180627696*10^{19}$. Adding it up to the earth's mass, it will be a (tM) of the earth-moon $=1.10028114711*10^{21}$ kg. Taking approximately a mean absolute temperature of 2000^0 Celsius, for both the earth and the moon as one body of mass, and multiplying it by their total mass, it will be their total heat power equal to: $2.20056229422*10^{24}$. Using distances from the Encyclopedia Americana on pages 11 and 535, I found the mean distance between the poles of the earth-moon and the sun to be approximately 149,538,928,746 meters, and the speed of the earth-moon per second in meters approximately 29,773. I counted the time for one revolution to be 31,558,150 seconds. With this shaped planetary system between the sun and the earth-moon, the situation will become similar to a classic electrical machine (see chapter 10). The whole thing will look like this: the sun with a nuclear reaction inside its mass is producing a thermodynamic energy released towards all its planets, and separately to each one of them. The total solar system is a thermal local equilibrium. Looking at the situation of the earth-moon and the sun, the energy released by the sun is a part of its total power and it will be supplied to the earth-moon by a powerful speed of light in the square or radiation which is a thermodynamic power that could symbolize a positive current in amperes per second or the power of the sun in watts per second. This power in watts is returned back to the sun by the kinetic energy (Ke) of the earth-moon but not instantly.

The heavy earth-moon as one body of mass will absorb part of the total radiation of energy from the sun every second, and this at the same instant will be returned back by its kinetic energy. At the end of one revolution of about 31,558,150 seconds the earth-moon will have returned back the total supplied energy from the sun, which has already been received up to this moment. This should be a portion of the total thermodynamic power of the sun. The rest of its radiation power will be retained for all of its 9 planets turning around the sun. This is happening because a potential difference of (HP-hp) is created between the sun and all the other planets in space simultaneously, including the earth-moon. This potential difference will generate a vacuum in space between the sun and the earth-moon disturbing the balance of the local thermal equilibrium. As a result an electro-motive force (EMF) between the two objects will be created, which is the forerunner of a force of attraction.

Simultaneously an (el-ch) from the earth-moon will oppose it for a moment, but in the end a continuous (FATT) from the sun will be established. The greater power of the sun will win the restrained effort and the limited resistance of the earth-moon and force them to be pulled around it. Here are more explanations with real numbers. The (FATT) from the sun is equal to: $(HP-hp)*c^2/(mDist)^2$

$$=(2.57032308098*10^{33}-2.20056229422*10^{24})*300,000,000^2/149538928746^2$$

$=1.03447903808*10^{28}$. [See Figure No.15 on page 102 about (mDist) and $2\pi r$]. A mighty force of attraction with a tremendous power of radiation and speed of light in the square from the sun towards the earth and the moon. That would be equivalent to a supplied current (I) in amperes per second. It will arrive at the earth-moon after the total power of the sun has been divided by the resistance of $(mDist)^2$ or the $(mRd)^2$ in ohms encountered in space. This mean distance in the square is equal to: $(1/2\pi)*(mRd)*2\pi*(mRd)=(mRd)^2=2.23618912106*10^{22}$ ohms. The initial thermodynamic power of the sun, before it started to travel in space as a speed of light in the square, and before it was divided by its resistance of $(mRd)^2$ was $(HP)*c^2$. This was equal to: $2.57032308098*10^{33}*300,000,000^2= 2.313290772288*10^{50}$. The earth-moon in one complete revolution around the sun will also have the same resistance of $(mRd)^2$. This is the resisting power of the earth-moon for 1 meter multiplied by the total traveling distance of $(2\pi R)$. It would be $(mRd)*(1/2\pi)*2\pi R$ (see chapter 14). If I divide $1 / (2\pi R)$ I get the revolutions of the earth-moon for 1 meter. And with $(1 / 2\pi)$ * R I get the resistance of the earth-moon in 1 meter of its traveling distance. In the next chapters you will find out the details of the FORCES OF ATTRACTION FOR MACROCOSMIC AND MICROCOSMIC PLANETARY SYSTEMS. Going further in this chapter now, the earth-moon being pulled around the sun will produce a negatively charged current (I), and develop a resistance by its kinetic energy. The time for 1 revolution of the earth is estimated approximately 31,558,150 seconds. The current (I) that has to be produced by the earth in one complete cycle is exactly the total thermodynamic power of the (FATT) received by its heavy mass in 31,558,150 seconds. The resistance developed by its (Ke) during the same time will be (el-ch) * 2π * (mRd) as it has been mentioned above. The negative resistance of the earth-moon and the positive resistance of the speed of light in the square are both the same because they are developed in the same line of the flowing current. The useful work of the earth-moon is the produced current (I) in amperes per second by their (Ke), and the useless one is the developed resistance. The two multiplied together in any one second, or in any one traveling meter will balance the power of the sun supplied to that second or up to that travelled meter.

The power of the sun in amperes and resistance per second are always re-turned back instantly by the (Ke) of the earth-moon. This all has to be done by ne-cessity. The earth's kinetic energy creating a negative current and a resistance at the same time during a complete cycle around the sun, has the purpose of filing up the vacuum created by the potential difference between these two objects of mass and balance the irregularity created in space. The heavy mass of the earth-moon and the long (mDist) from the sun will slow down their speed per second to complete a whole cycle and match the lightning magnitude of the (FATT) expressed in speed of light in the square (see Chapter 14). I recall the relation to convert meters to revolutions is: meters/$2\pi R$=revolutions. For example: 1 meter divided by the ($2\pi R$) of the earth-moon or 939580799950 meters is equal to: $1.06430442178*10^{-12}$ revolutions and 29,773 meters will correspond to $3.16875355495*10^{-8}$ revolutions. The same will apply for any electron turning around its nucleus. It is mentioned in the previous chapter THAT THESE REVOLUTIONS FOR THE EARTH OR THE ELECTRON TO COVER ONE METER DISTANCE AND THE TIME IN SECONDS TO TRAVEL THIS DISTANCE IS MUCH DIFFERENT IN ANY ONE PLANETARY SYSTEM EXISTING IN THE UNIVERSE; including the earth-moon's and the electron's. For the earth-moon to cover 1 meter distance it will have to complete $1/2\pi R$= $1.06430442178*10^{-12}$ revolutions. And the time in seconds to do it is: 1/29,773 which is $3.358747588*10^{-5}$ seconds. For the electron in order to cover that same distance of 1 meter it will need $1.59154943092*10^{17}$ revolutions and the time that it takes is 1 sec-ond. When the (mRd) becomes extremely small the (FATT) will become too strong and the speed of the revolutions are going arbitrarily high. IN THIS CASE THE STRONG NUCLEAR FORCE GETS THE HIGHEST MAGNITUDE BETWEEN THE NUCLEUS AND THE ELECTRON. THIS IS BECAUSE OF A VERY HIGH FORCE OF ATTRACTION UPON THE ELECTRON, AND AN EXTREMELY SMALL DISTANCE AND RESISTANCE BETWEEN THEMSELVES TO AN INFINITESIMAL DEGREE. The number of revolutions per second and the num-ber of the amperes per second produced by the electron would be very high indeed, and this electron must use only 1^0Celsius above zero per second. The difference in (HP-hp) between (e$^-$) and (n) is too small but when it is multiplied by the (c^2), and divided by the very low resistance of (mDist)^2 it will produce a tremendous force of attraction equal to: $3*10^{52}$. The great number of the electric charges crossed per second by the (e$^-$) would make it produce the highest number of amperes per second, which then the amperes multiplied by an extremely low resistance now, will be re-turned back to nucleus and equalize its (HP). The same thing would happen to the earth-moon and the sun. If I know the (FATT) supplied and the (mRd) for these celestial bodies of mass I might be able to calculate accurately their amperes and re-sistance produced by them in meters for any period of time in seconds that they have travelled in their elliptical orbit around the sun. For instance in 29,773 meters or one second, traveling distance on the line of their elliptical orbit around the sun, the amperes produced at that point would be: $(29,773/939580799950) * 1.03447903808 * 10^{28}$, or $1/31558150 * 1.03447903808 * 10^{28}$, the results will be the same in both equations and equal to: $3.27800912944 * 10^{20}$. Following the same procedure the resistance at the same time should be: $29,773/2\pi R$ revolutions or $29,773/939580799950$ multiplied by the total resistance for the $(2\pi R)$ distance, which is the $(mRd)^2 = 2.23618912106 * 10^{22}$; and the answer would be $7.08593222689 * 10^{14}$. What I am actually doing is transforming the distance in meters travelled by the earth-moon to revolutions, at the point where I want to find out how many amperes have been produced, or how much resistance has been developed? And then take the answer and multiply by the total amperes produced in one cycle or the total resistance. For the case above, I will get:

$3.16875355495 * 10^{-8}$ revolutions $*1.03447903808 * 10^{28} = 3.27800912944 * 10^{20}$ amperes, and for the resistance I will get:

$3.16875355495 * 10^{-8} * 2.23618912106 * 10^{22} = 7.08593222668 * 10^{14}$ ohms. Amperes and resistance multiplied together will equal to: $2.32277505297 * 10^{35}$. And that is work done for one second by the earth-moon. For two seconds there would be two times the amperes produced in one second, and two times the resistance produced in one second.

$(2 * 3.27800912934 * 10^{20} = 6.55601825868 * 10^{20})$ and

$(2 * 7.08593222668 * 10^{14} = 1.41718644534 * 10^{15})$. Multiplying both of them together I get $9.29110021158 * 10^{35}$. The same results could be obtained if I multiply the total work that has been accomplished by the earth-moon in one second by the number of seconds in the square. Or if I multiply the work done by the earth-moon in one second by the total seconds for one revolution in the square and then I get the total work in watts being completed by the earth to be equal to the total power of the sun, that has been supplied to the earth-moon to complete one revolution in this planetary system. It is:

$(HP-hp) * c^2 = I * (mDist)^2$, or it is: $31558150^2 * 2.32277505289 * 10^{35} =$ the total sun's power equal to: $2.31329077078 * 10^{50}$.

The power of the sun in percentage can be shown by the earth-moon, at any period of time in seconds. This would be a fraction of the total thermodynamic power in watts that has been supplied by the sun to the earth-moon at that particular moment.

For one second it will be:

$(2.32277505289 * 10^{35}) / (2.3132907788 * 10^{50}) = 1.00409990829 * 10^{-15}$ of the total power of the sun. For two seconds it should be:

$(2^2 * 2.32277505289 * 10^{35}) / (2.31329077288 * 10^{50}) = 4.01639963316 * 10^{-15}$. And for 31558150 seconds it is:

$(31558150^2 * 2.32277505288 * 10^{35}) / (2.31329077288 * 10^{50}) = 1$. The unit one will represent the total part of the power in **WATTS** produced by the sun, which is destined to supply the earth-moon.

THE SUN, THE EARTH AND THE MOON

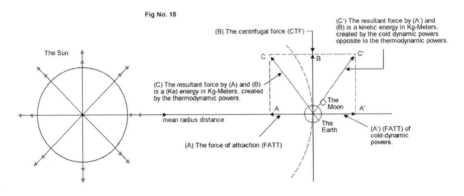

Fig No. 15

(B) The centrifugal force (CTF)

(C') The resultant force by (A') and (B) is a kinetic energy in Kg-Meters, created by the cold dynamic powers opposite to the thermodynamic powers.

The Sun

(C) The resultant force by (A) and (B) is a (Ke) energy in Kg-Meters, created by the thermodynamic powers.

mean radius distance

(A) The force of attraction (FATT)

The Moon

The Earth

(A') (FATT) of cold-dynamic powers.

The vector "A" is the "Force of Attraction"(FATT) or the centripetal force (CPF). The vector "B" is the centrifugal force (CTF). The vector "C" is the resultant force created by the "A" and "B". It is a powerful kinetic energy by the earth and the moon, expressed in kilogram-meters that will balance the tremendous force of attraction exerted by the huge thermodynamic power of the sun at any period of time during a cycle by the earth-moon around the sun. The earth-moon is being negatively charged, and of course they will produce a negative current while they are moving around the sun. The force of attraction (A) imposed to the earth-moon, is developed by the thermodynamic power of the sun, and may also be named a centripetal force (CPF), which acts as a counterpoise to the centrifugal force (CTF), drawing the earth towards the center of the sun. It is acting like a rope, holding up the earth and not allowing it to fly away from the sun. See the next chapter about the force of attraction. The centrifugal force will be tending to fly off the earth at any point of its motion to the direction of the tangent of the curve of its elliptical orbit. This is a force disputed by both the thermodynamic and the cold-dynamic powers. The resultant (C) is the outcome of the other two forces of (A) and (B). It is a kinetic force moving the earth around the sun.

CHAPTER 16

FORCE OF ATTRACTION OF THE MACROCOSMIC PLANETARY SYSTEMS IN THE OPEN SPACE-THE EQUATION OF THE GRAND UNIFIED THEORIES

THE FORCE OF **ATTRACTION (FATT)** either on the side of the thermodynamics or the cold-dynamics is a phenomenon displayed in all the thermal local and distant equilibrium in the whole space of the universe with its total planetary systems. The force of attraction is developed by a potential difference of hot and cold power between two objects or particles of mass anywhere in space. This potential difference is the (HP-hp) of the objects and it has to be multiplied by the speed of light in the square, and divided by the mean distance in the square; all measured in meters. It is (FATT) = (HP-hp) * c^2 / $(mDist)^2$. As it is seen in this equation the factor $(c)^2$ which is the speed of light in the square is included, to make that potential difference very strong indeed.

It is an equation similar to the classic electrical one (E=I*R). The (HP-hp) represents the (E) and the $(c)^2$ represents the current (I) in amperes which is a kinetic energy of the speed of light in the square. It constitutes a primary and positively charged current in space. The source is always a body of mass or a particle of mass in space. A very hot energetic body of mass like the stars or galaxies positively charged, which automatically creates a weaker and secondary body of mass negatively charged, pulled around to resist and counterbalance the power of the first one. The pulled object or particle with its (Ke) will produce a negatively charged current (I) in amperes expressed in kilogram meters and a resistance $(mDist)^2$, which multiplied together at any moment will equalize the corresponding power supplied by the positively charged object at that same moment. It is similar to the classic electrical circuits, where the total (I * R) equals the (E). The (FATT) here is more extensively explained, as it is very important in the cosmic physics.

As soon as a potential difference in space is created between stronger and weaker in heat power masses automatically an (EMF) will be created, and an electric charge by the pulled object is developed. The intrinsic properties of our temperature hot and cold, are the responsible factors, for all these things happening all over the space in the universe. **The very important EQUATION of the force of attraction I believe is the same equation for all four forces in our universe; the strong nuclear power, the weak nuclear power, the electromagnetic and the gravitational ones.** I strongly believe that this equation solves the problem of the strong nuclear force and the gravitational one, as well as the other two forces mentioned above.

There is no room in space without the existence of a force of attraction, whether it is developed in a macrocosmic distance, which will appear as a gravitational force or in a microcosmic distance, which will be seen in all atoms as the strong nuclear force. For both these forces and all electro-magnetic and weak ones in the whole universe, the force of attraction is responsible. The higher the potential difference between the hot and the cold power is, then the higher the force of attraction will be. This hot and cold power exists anywhere in the universe and at any cosmic time. Even when the universe is "DEAD" in an absolute zero degree of temperature because there are still 14/27 parts of cold-dynamic and 13/27 of thermodynamic ones in the pre-existing mass to react.

The pre-existing mass in almost a solidarity form won't endure this situation too long, before another big bang occurs. If this degree of absolute zero temperature will appear somewhere in the interstellar space of a live universe, it rather would be a black hole as named by the astrophysicists. This I think is a spot inside our global universe, which is communicating, with the extreme outskirts of the universe, where rigid coldness of an absolute zero is prevailing. It is a bottomless hole likely with a huge potential absorbing power. There could be developed as a result, a cold-dynamic force of attraction similar to those created in the first picoseconds of the big bang. It could have such strength that it would be able to absorb anything nearby instantly, including strong beams of light or even whole galaxies.

This is an inversion of a thermodynamic (FATT) to a cold-dynamic one. This force of attraction, when developed has one of its factors the most dynamic absolute zero degree equivalent to the top highest degree of temperature. In a live universe now, there is an average degree of temperature in space separating the two REACTING almighty powers of cold-dynamics and thermodynamics. This is the line of the zero degrees of Celsius temperature. From this point and up, the thermodynamics will dominate the space, where below this point the cold-dynamics will take over. When the thermodynamics are in the throne, there is only one force predominating in the whole space and this is the thermodynamic force of attraction. Most of the time it is a nuclear heat power source (like a star or galaxy), that its strength or luminosity multiplied by the speed of light in the square would make it very powerful, and cause other weaker masses in the universe to move around and develop other slower kinetic energies in kilogram- meters.

All these kinetic masses are negatively charged and by moving around they will counter balance the tremendous positively charged current of the speed of light in the square of the stars or galaxies at any moment as it has been mentioned many times before. No other force in a live universe than those forces of attraction could be of a higher magnitude and be of that much importance. The pulled object around will have the purpose of producing a current (I) in amperes and a resistance, to fill up the vacuum and the irregularity created in space every second, and bring about a balance in order to make the universe live for billions of years. The same circumstances as they have been explained in chapters 14 and 15. The positively charged light emission or radiation of temperature in the square meters is moving through the space, in a radiant and /or circular, elliptical, or spiral direction within the globular universe, to attract and develop forces of attraction. (See Figure No.16). It is unbelievable to watch a huge planet like earth get pulled around by the sun and develop such incredible high speeds per second in space, and not be aware of what causes this. Under zero degree of Celsius temperature, the cold-dynamic powers will take the upper hand and create cold-dynamic forces of attraction absorbing the lower thermodynamic powers wherever they will appear, but not being able to shape planetary systems with stars and galaxies because there would not be high nuclear sources of thermodynamic masses composed of stars or galaxies to react any more. When the average temperature in space reaches the zero degree of Celsius, the period of time that is needed after this to bring the universe to an end I believe is confined. A whole universe that has lived so many billions of earthly years will come to an end in a matter of very few degrees of difference, from 0^0 to -273.16^0 Celsius.

It will be dead. All of this said above, could form and set a famous equation. This is the equation of the force of attraction (FATT). It is the heat power of the positively charged object or particle of mass minus the heat power of the negatively charged object or particle of mass, or it is the higher powered negatively charged cold mass minus the lower powered negatively charged mass, multiplied by the speed of light in the square, and divided by the mean distance in the square between their poles in meters. The equation for the thermodynamics is: (FATT) = (Hp-hp) * c^2 / (mDist)^2, and for the cold-dynamics it is: (FATT) = (CP-cp) * c^2 / (mDist)^2. (CP) and (cp) are the positive cold power and the negative cold power. The powerful speed of light emission or radiation of temperature in the square when the universe is active and expanding is decreasing in power, to the square of the (Dist) in meters. I wondered many times how far for instance a planet could be in space so that it can be attracted from the sun? I believe that the thermodynamic power of the sun can only attract a planet up to a limited distance in meters, equal to the square root of the thermodynamic power of the sun, which is

(HP) * (c²) as I have found out.

Its force of attraction could have no energetic power beyond that point in space. If the sun's power for instance is: 2.31329077288 * 10⁵⁰, its square root shall be 1.52095061487 * 10²⁵ meters. Any celestial object being positioned in that distance shall not be able to move and turn around the sun I believe. It will stay idle and cold in space. It might be pulled out by other superior power sources in the interstellar space. That is why humans made satellites, purposely located at a certain distance above the earth's center in space so that they could remain idle and not be attracted from the (FATT) of the earth. The (Dist) between these satellites and the center of the earth in meters I think should be equal to the square root of (HP) * c² of the earth or the square root of

(mT * earth's mass * c²).

From this equation I MAY BE ABLE TO FIND OUT ACCURATELY THE MEAN TEMPERATURE, THE TOTAL MASS OF THE EARTH, and the (mDist) in meters from the centre of the earth to the satellite if I only knew two of these factors. In the earliest cosmic times, the pulled objects had an almost circular orbit in their planetary systems. The reason was that there wasn't much reaction from the cold-dynamic powers. But later in the middle cosmic times this orbit had changed to an elliptical one because there was more reaction from the cold powers in space. And while the universe is approaching its own end, all the pulled objects in their planetary systems are receding faster from their superior positively charged stars and galaxies, changing their orbits to a spiral direction. I think this is happening because the cold-dynamic powers in space are increasing, and clusters of galaxies are traveling closer to the outermost layers of the sphere of our universe, where an almost 100% crystallized solidity of an absolute zero degree of temperature is existing. And therefore higher (EMF's), will develop strong forces of attraction between these galaxies and the lowest rigid temperatures.

When these stars and galaxies are approaching the extreme borders of the global universe, they will become an easier prey for these superior powers of cold-dynamics. Any object of mass moving through will be easily sucked in by a powerful (FATT), between the galaxies and the huge mass of blackness and rigid coldness that would form the cold-dynamic powers. The universe will start to shrink. The outermost layers of the sphere and most parts of it would rapidly change to a hell or Hades of absolute darkness taking the shape of a metaphysical world. This situation will precipitate the end of the universe, and the remaining masses in space might disappear more easily and be transformed into a PRE-EXISTING MASS OF COLD AND HEAT POWER. This dark mass will form a shape of a sphere in absolute solidity. It would look like an implacable, inexorable and impenetrable WALL of consolidated MASS of temperature, unyielding and unbending to any means of power in the universe. In situations like this, isolated whole aggregation of stars and galaxies may disappear and get lost.

Reading newspapers and magazines often, I am pretty sure I remember astrologers saying that they observed galaxies, which are receding much faster from other galaxies in far distances from here. Also I believe I remember reading that recent leading theorists called this outermost layers of coldness and blackness, "a massive dark matter". In my opinion they were very close to the reality. I strongly agree that it is a massive dark thing as the theorists said, but it is not matter. I believe that this dark thing is a dualism of temperature in belief of two fundamental existences and the belief in two antagonistic supernatural beings, the one good and the other evil. That is what is there. And that degree of absolute zero temperature is unyielding and unbending to any means of power in the universe. It is something, which is nothing, but this nothing is coming out from something. Remember that this dualism will boast and it will command the whole universe.

The force of attraction will be nothing less, nothing more, than that which I thing was for centuries entangling and puzzling many theorists and scientists of the world. The force of attraction also holds the secret of the link between the thermodynamic and cold-dynamic powers. This force of attraction has the two existing planetary systems working in harmony for billions of years, and it holds the total masses of stars and galaxies in the universe together for long periods of cosmic times and does not let them fly apart when it is in action. All the stages of local and distant thermal equilibrium in a live universe are based on the cosmic law of the force of attraction. The scientists' aspect that the gravitational forces are to be the weaker forces along with the electro-magnetic ones, in my opinion is not right any more. All those four existing forces in the universe originate from one only cause. This cause is the POTENTIAL DIFFERENCE BETWEEN HOT AND COLD MULTIPLIED BY THE SPEED OF LIGHT IN THE SQUARE. And the solution for all of them is in one equation. This is the equation of the force of attraction of the thermodynamics and the cold-dynamics. And I strongly believe that the ONE EQUATION OF THE FORCE OF ATTRACTION WILL SOLVE THE PROBLEM OF THE STRONG AND WEAK NUCLEAR FORCES, AND THE GRAVITATIONAL AND ELECTRO – MAGNETIC ONES. THIS MEANS THAT THIS EQUATION IS THAT OF THE GRAND UNIFIED THEORIES. The force of attraction emanates from the existence of a duality of temperature in one entity. The whole universe also could be pictured as the equation of: $0 = A / \infty$ and $A = 0 * \infty$ and also $\infty = A / 0$. Where (A) constitutes any quantity or capacity of whatever immaterial substance, including active or inactive objects of mass, volume of space, light and dark, or any abstract kind of number. The absolute zero (0) will have the capacity in itself to hold a whole universe. "Infinity" will include mass, volume of space, and numbers that cannot be counted. See chapter 2. Also see Figure No.16.

CHAPTER 17

FORCE OF ATTRACTION IN THE MICROCOSMIC PLANETARY SYSTEMS OF THE ATOMS

THERE IS ANOTHER situation developed here with the force of attraction in the depths of the micro-cosmic domain of the atoms. Here the intrinsic powers of the leptons of heat and cold, are dominating the issue as you may see next how the atoms are functioning in their smallest ever microscopic planetary systems. The nuclei of the atoms consisting of protons; or protons and neutrons. Each one of the atoms of course having electrons spinning around their nuclei. I think I know how their force of attraction is developed, and how they are functioning like that for billions of years. Let me start from the protium atom first. The nucleus of the proton consists of 2,754 positively charged fractional thirds and 2,754 negatively charged fractional thirds. These all are fractional thirds of thermodynamic and cold-dynamic power of condensed temperature intrinsically, and counteracting each other, leaving a zero heat and cold power in the nucleus itself. Any outside temperature in the neighborhood of the atom, would not influence its status of neutrality. Let it be for instance a neighboring average temperature in space of 21^0 Celsius above zero. With these degrees of temperature added, I multiply the number of the positively charged thirds, as well as the negatively charged ones separately. The result is: (21 * 2754) = 57,834 heat power of the thermodynamics and: 21 * (-2754) = -57,834 cold power of the cold-dynamics, which of course will counteract each other, leaving a zero heat plus cold power again. The same results will be obtained with any other external temperature in the vicinity of this atom, being either below or above zero Celsius. If it is for instance 21 degrees below zero Celsius, the results will be the same as before: -21 * 2754 = -57,834 for the positively charged fractional thirds, and:

-21 * (-2754) = 57,834 for the negatively charged fractional thirds.

Adding them up I will get a zero again. That would rather be normally admitted as I have explained it before in chapter 6 describing the leptons and their construction etc. But the nucleus of the proton is the heat power source and it is supposed to be positively charged in order to develop a force of attraction to the negatively charged electron and thus to create a planetary system in the atom and make the electron spin around. In other words there must be a difference in temperature, which will produce a difference in the heat power between the nucleus and the electron for this purpose. Here I recall

(HP = mass * mT). I WILL SAY THAT THERE IS ALWAYS A DIFFERENCE IN TEMPERATURE AND CONSEQUENTLY A DIFFERENCE IN HEAT POWER BETWEEN THE NUCLEUS AND THE ELECTRON! And as an example, let me assume that the average temperature in the vicinity of the atom is 1 degree Celsius above zero. Here I have a difference in heat power of (1/3). But how can this all happen? Human wisdom has prevailed again, over the mysterious problems and the tricks of nature. I have the opinion that my hard work allowed me to succeed and make it feasible to surface that suspicious trick of nature. Here again it seems that God plays with dice. I think that Einstein had said that God doesn't.

The difference in heat power between the nucleus and the electron shall be 1/3. That is: 0 - (-1/3) = 1/3! This is happening because another situation has been created between the (n) and (e⁻) now. When the external temperature in the neighborhood of the atom is shifted to a higher or to a lower degree, the nucleus of the simplest protium atom of the hydrogen isotope will remain in the same neutral status as I said above, but the electron because of a different construction in itself will change its status each time the external temperature is changing. Every time there is a lower or a higher external temperature degree in the near by space of the (e⁻), its heat power will be different from the initial one it had possessed before. Always in these circumstances the electron is influenced by this changing temperature; but not the nucleus of the protium atom though which is remaining in the same neutrality (zero charge), because its total fractional thirds are 50% cold and 50% heat power so that any change in temperature is equally distributed in to the (+) and (-) fractional thirds, without any influence to the status of the nucleus except whenever there is a mixture of protons and neutrons involved. I will talk about it a little later. The electron (e⁻) turning around the nucleus is consisted of two thirds in cold-dynamic power, and one-third in thermodynamic power (see chapter 6), and this will make the difference of course. Because if I assume that the 21^0 Celsius above zero remains the same again in our example, the situation will be different for the whole atom in both cases. The difference in temperature and heat power simultaneously between the nucleus and the electron will not be the same any more. With 21^0 Celsius above zero the electron will have: 21 * (-2/3) cold power and 21 * (1/3) heat power in itself or: (-14) + (7) = (-7) remaining cold power. Theses values of temperature will change not only the difference of the (HP-hp) between the (n) and the (e⁻), but the nucleus power of (HP-hp) * c^2 as well. Also changed will be the more important factor of the (mDist) between the (n) and the (e⁻), which in general will influence all other factors including the amperes and resistance.

In atomic planetary systems the work being done by the (e⁻) defers from that in the external planetary systems. Here the (e⁻) using its (Ke) would always need one degree of temperature or more above zero to produce different amperes and resistance in one second. The (FATT) imposed on the (e⁻), would look the same as $(3 * 10^{52})$ as we will see it later, but the amperes and the resistance of the (mDist)^2 I think would always be depended on the average temperature around the atom. The number of amperes produced naturally will be depended on the higher or lower number of revolutions and the crossed electric charges per second by the (e⁻). These two factors would be changed drastically every time there is a change in temperature around the atom, which at the same time will influence the (mDist) between the (n) and the (e⁻). The less or more amperes produced in this case multiplied by the resistance developed by the (e⁻) every second, would be returned back to the nucleus. If for instance the (mDist) between the (n) and the (e⁻) is twice as much more as that in the protium atom because of a changed temperature, the revolutions per second as well as the number of electric charges would be reduced to half per second. In that case the amperes produced would be: 1/2 * 1/2 = 1/4, and the resistance developed would be four times higher than what it was before. And

4 * 1/4 = (1^0) of thermodynamic temperature. The (e⁻) will consume only 1^0 Celsius temperature every second.

But if the (e⁻) needs only one degree Celsius temperature to produce amperes and resistance, where would the other degrees of extra temperature go? This and many other details are discussed later in the next chapter. One may ask how did I find out about all these important issues concerning cosmic physics? Well, I could not tolerate that an (e⁻) turning around the nucleus would hold the same (mDist) while the temperature in the vicinity could vary from one to thousands, or even to millions of degrees of temperature above zero Celsius. I knew that with a higher or lower temperature around the atom, the electron (e⁻) could move a bit further or a little closer to the (n). I was curious to find out how much it would move. I was thinking until I reached the point that there was a relation between a standard (mDist) of the (e⁻), and the square root of the changed temperature in degrees of Celsius around the atom. What is a standard (mDist)? This is a fixed (mDist) between any (n) and any (e⁻) in any atom, when the average temperature around it is 1^0 above zero Celsius. I was very happy to have discovered that this temperature MUST BE one degree above zero Celsius, and happier yet that this (mDist) between (n) and the (e⁻) in the protium atom should be $(1 * 10^{-18})$ meters, because in that temperature around this atom there is no expansion or contraction at all. You will read more details about it later. And the next thing was, what should be the new (mDist) between the electrons from their nuclei each time the temperature changes around them. I was very happy that I found that too. Indeed, in order to find out the (mDist) of an (e⁻)

any time that the temperature around the atom changes, all I have to do is multiply the square root of the changed temperature by the standard mean distance of this particular electron. Finally with a known changing temperature each time, I would be able to estimate accurately the new (mDist) between the (n) and the (e⁻) in any atom, and of course the other factors like the difference in (HP-hp), the nucleus power of (HP-hp) * c², and the (FATT). Let me now examine the force of attraction upon the electron in a protium, and then in a deuterium atom at an average temperature of 1^0 Celsius above zero. In this temperature the standard mean distance between the (n) and the (e⁻) is equal to: 0.000000000000000001 meters, or: $1 * (1 * 10^{-18})$ meters. It is the square root of 1^0 Celsius multiplied by the standard (mDist) of the electron. The (HP-hp) is:

0 - (1 * -2/3 + 1 * 1/3) = .333333333333, and if this is multiplied by the speed of light in the square of (c²) I will get the (n) power of $(HP-hp) * c^2 = 3 * 10^{16}$. If it is divided now by the (mDist)², it will become a (FATT), which will be expressed in a speed of light in the square, measured in amperes per second and being imposed to the (e⁻). The (e⁻) with its (Ke) in kilogram-meters will in turn produce a negatively charged current in amperes, which multiplied by the resistance developed (it is the same resistance between the nucleus and the electron) will be returned back to the (n) power. It is an equation of:

(HP-hp) * c² = (FATT) * resistance. But when the power of the nucleus is divided by a decimal resistance (meters in the square under the unit one) the (FATT) imposed to the (e⁻) will become tremendously high and equal to: $3 * 10^{16} / (1 * 10^{-18})^2$ $= 3 * 10^{52}$; and the (e⁻) will produce amperes one after the other every second. This number of amperes multiplied now by the same very low resistance will equalize the nucleus power. It is:

$(3 * 10^{16}) = (3 * 10^{52}) * (1 * 10^{-18})^2$. The resistance produced in the line every second is common for both the positive (n) and the negative (e⁻). The nucleus power has to be divided by this low number under the unit one resistance, in order to produce a huge force of attraction in amperes per second and the same resistance has to be multiplied by this number of amperes produced by the (e⁻) in order to return back the same supplied power by the (n). The electron's kinetic energy in kilogram-meters will balance the (FATT) supplied by the nucleus at any moment. You cannot imagine that when the average temperature around the protium atom is 1^0 Celsius the electron turning around its nucleus will produce a useful work of one only ampere in every one second. In higher temperatures there would be higher mean distances created between the (n) and the (e⁻); they are going to develop higher resistances and less amperes but not a lower force of attraction, as this will remain the same. See the next chapter for more details. When this nucleus is mixed up with protons and neutrons, the situation will not be the same any more. It will change but just a bit.

Following the same procedure for a deuterium atom, I will also count its (FATT). Let me assume again that the temperature around the atom is 1^0 Celsius. In the nucleus there will be 1,836 + 1,839 = 3,675 leptons, or 11,025 fractional thirds, of which 5,512 of those will be positively charged and 5,513 negatively charged (see chapter 7). The nucleus negative power is equal to:

5,513 / 11,025 = 0.500045351474, and its positive power is equal to:

5,512 / 11,025 = 0.499954648526. Their difference is slightly on the negative side. It is:

-.000090702948. Multiplied by the 1^0 Celsius average temperature, I will get the same difference. And this is the (HP) of the nucleus. The (FATT) is: (HP-hp) * c^2 / $(mRd)^2$. That would be: -.000090702948 - (1 * -2/3 + 1 * 1/3) * $300,000,000^2$ / (1 * $10^{-18})^2$ = 2.99918367347 * 10^{52}. Slightly different from the (FATT) of (3 * 10^{52}) in the protium atom.

And this is because the nucleus is mixed up with only one neutron. In other heavier atoms with more protons and neutrons in their nuclei, and more electrons turning around you may find again very little difference in the forces of attraction from that of

(3 * 10^{52}). The heaviest atoms have about 3 neutrons for every 2 protons in their nuclei, while in lighter elements the nuclei have about equal number of protons and neutrons. You may find them in the periodic table of elements, with the names of their discoverers; uranium for example has atomic weight of 238,made up from 92 protons and 146 neutrons. I would like to mention here that there are many details in this chapter that need to be discussed further, therefore the following chapter is more extensive and it will explain all of these important matters, which in my opinion constitute fundamental issues in cosmic physics.

Fig No.16

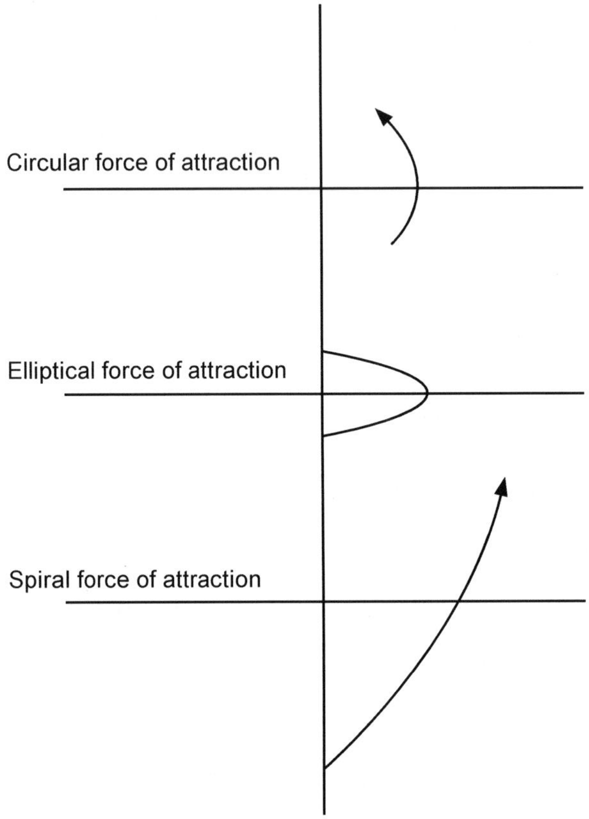

Circular force of attraction

Elliptical force of attraction

Spiral force of attraction

CHAPTER 18

THE NUCLEI AND THE ELECTRONS IN THE ATOMS, AND THEIR IDEAL PERFORMANCE AS MICROCOSMIC PLANETARY SYSTEMS

MY CHOICE ALWAYS was to select the planetary system of the protium atom, comparing it to the external ones in order to learn the secrets of nature, and this paid off since I found out fundamental issues of cosmological dimensions. This protium atom can be used as a unit to measure, compare and find out factors for other heavier kinetic masses in space, like for instance the earth and its revolutions, forces of attraction, and electric charges, etc. This smallest and plainest planetary system in our cosmos has some unique and exceptional functioning performances that will make it the ideal planetary system of a local thermal equilibrium in the whole misleading world of our universe. I have mentioned before in chapter 14 that the (mDist) from the nucleus to the electron in the protium atom is: $1 * 10^{-18}$ meters, its (el-ch) is: $1.59154943092 * 10^{-19}$, and the number of revolutions per second is: $1.59154943092 * 10^{17}$. These revolutions per second were made by the (e^-) to cross $2\pi * 10^{18}$ electric charges in one second, to produce the power of one watt.

All these things above of course would be so if the average temperature around the atom could stay at 1^0 above zero Celsius. In lower or higher degrees of temperature, there are different situations created and the above-mentioned numbers could not necessarily hold true. I have explained much about these situations in the previous chapters. The nucleus and the electron will follow almost similar functioning performance as that related to the sun and the earth detailed in chapter 15. I will try here to explain it much better, of how perfectly this planetary system is functioning in its atom. We should know by now that the property of the heat power ratio in the pre-existing mass (which mass is nothing else than hot and cold condensed temperature) is 13/27; and this part is the only one energetic factor that is going to influence the total mass in the whole space in the universe, so that it could be expanded or contracted. **THIS THING ALONE** has made me very deeply concerned. The thermodynamic temperature starts from zero degrees of Celsius and up. From that point you will see the (e^-) start to move in its microcosmic planetary system, and in much higher temperatures the stars and galaxies will be following, shaping and developing the external planetary systems. The circular electro mechanics at the same time will be applied. From zero Celsius temperature and down below there is a motionless electron. More about it is explained later in this chapter.

The zero Celsius is the middle line of thermodynamic and cold-dynamic powers. From zero Celsius and up, the (e-) will start to move around its nucleus. In zero Celsius degrees of temperature the electrons in the atoms are stuck together with the nuclei. From zero Celsius temperature and down to the absolute zero degree, the strong nuclear force between the (n) and (e-) is infinite! An infinitesimal amount of thermodynamic temperature above the zero Celsius will make the electrons move around the nuclei. At that moment the (mDist) between the nuclei and the electrons would be the smallest ever. The nucleus of the atom in this low temperature will have the very smallest power. The resistance of the (e-) would be at its lowest point. The (FATT), imposed to the (e-) will be the same as always $(3 * 10^{52})$ but what happens now? The electron will increase its revolutions and consequently it will cross more electric charges to produce more amperes per second, and I think we may observe here what our particle physicists have been looking for, for a long time. I sure believe that the more amperes produced now by the electron will make the nuclear force get stronger and this is how the nuclear force is developed higher in the atoms. In extraordinary small (mDist) between the (n) and the

(e-), the nuclear force will be getting arbitrarily high up to the point where the (e-) contacts the nucleus and the (FATT) becomes infinite. In contrast of a very low temperature around the atom the (e-) will consume 1^0 Celsius temperature per second again to produce more amperes and less resistance, which multiplied together every second would equal the nucleus power.

The force of attraction is expressed as a speed of light in the square before it is imposed to the electron. When the speed of light contacts the (e-) it will change into a positive current. Whether the (e-) is turning closer around its nucleus or further from it, the (FATT) is always the same and equal to $3 * 10^{52}$. This is because there is a relationship between the surrounding temperature, the power of the nucleus upon the (e-), and the (mDist) between the (n) and the (e-); and because of a change in the (mDist) in meters from the (n) to the (e-) all other factors changed too. When the power of the (n) is indicated as $(HP-hp) * c^2$ it will be shown by the (e-) as $(FATT) * (mDist)^2$. And when the power of the (n) is indicated as $(HP-hp) * c^2 / (mDist)^2$ it would be shown by the (e-) as a plain (FATT). It should be as: $(HP-hp) * c^2 = (FATT) * resistance$, or

$(HP-hp) * c^2 / (mDist)^2 = (FATT)$. When the thermodynamic temperature around the protium atom is 1^0 Celsius, this planetary system has an ideal performance in the space of the universe. It will produce a current of (1) ampere in one second, and develop a resistance of (1) ohm. But how do we find 1 ampere produced and 1 ohm of resistance developed by the electron turning around its nucleus? As you may see it explained later, when the thermodynamic temperature in the neighborhood of the atom is 1^0 Celsius the (mDist) between the (n) and the (e⁻) is: .00000000000000001 meters. The resistance or the (mDist)² is: .00000000000000001² ohms and the power of the nucleus is:

(HP-hp) * c^2 or 0 - (-1/3) * 300000000² = 3 * 10^{16}. This time the (n) consuming the above 1^0 Celsius thermodynamic temperature will come up with a power of (3 * 10^{16}); exactly what it needs to pull the (e⁻) with such a force so that it will complete the right revolutions per second, cross the right electric charges, and then it will produce (1) ampere and develop 1 ohm resistance. On the other hand the (e⁻) will have to consume 1^0 Celsius of thermodynamic temperature per second and this is independently of what amount of temperature is consumed by the nucleus at the same time. From 1^0Celsius and higher the amperes produced by the (e⁻) will be fractional and decreased as the temperature goes higher. The resistance would be integer numbers and increasing with higher temperatures. The opposite will be observed in the temperature from 1^0Celsius and lower. In the next pages you may find all the details about everything concerning this protium atom with the changing temperatures in space. When the (n) in the protium atom (average temperature around 1^0 Celsius) is divided by the negative resistance of

(1 * 10^{-36}) it would accurately increase its power from (3 * 10^{16}) to (3 * 10^{52}).

This (3 * 10^{52}) is the force of attraction developed and conveyed to the (e⁻). This supplied power by the nucleus simultaneously will give a power of (Ke) energy to the

(e⁻) so that it could turn around the nucleus with the right number of revolutions per second in order to cross (2π * 10^{18}) electric charges and produce 1 watt per second. The resistance developed by the (e⁻) would also be (1 * 10^{-36}) or (1 * 10^{-18})². This resistance multiplied by the powerful (FATT) inflicted to the (e⁻) will equal the (n) power and the power will be returned back to the (n). More clearly when the positively nucleus power is divided by a negative number of resistance, in this case the (1 * 10^{-36}) in meters between the nucleus and the electron will add to the nucleus power. It would look like this:

(3 * 10^{16}) / (1 * 10^{-36}) = (3 * 10^{52}). When this (FATT) now is multiplied by the same negative resistance of (1 * 10^{-36}) it will make: (3 * 10^{52}) * (1 * 10^{-36}) = (3 * 10^{16}). For the resistance now, I have: (1 * 10^{36}) * (1 * 10^{-36}) = 1 ohm. The resistance of the nucleus is positive. The resistance of the electron is negative. When the average temperature around the atom goes 4^0 Celsius, the (mDist) between (n) and (e⁻) will come up to (2 * 10^{-18}) which will be two times higher, and the resistance will be four times higher to (4 * 10^{-36}). See how it is explained in the next pages. The nucleus power will

increase to four times more: $0 - (4 * -2/3 + 4 * 1/3) *c^2 = 1.2 * 10^{17}$. When this happens the (FATT) would still show ($3 * 10^{52}$) but this (FATT) will render 1/4 amperes and 4/1 resistance. You may remark that the produced amperes would be the reciprocal of the changed temperature around the atom. In this case the revolutions per second and the crossed electric charges by the (e⁻) will be half than what they were in the protium atom when the temperature around was 1^0 Celsius. To find how we get 1/4 amperes, we multiply 1/2 revolutions per second by 1/2 crossed electric charges by this (e⁻) compared to that in the protium atom activated at 1^0 Celsius temperature. If the temperature around the protium atom reaches 16^0 Celsius, the produced amperes by the (e⁻) would be 1/16 and the resistance 16. The

(e⁻) is borrowing 1^0Celsius thermodynamic temperature every second to do all this by its (Ke). The temperature would always remain the main factor in estimating every time the new mean distance between the nucleus and the electron, and the power of the nucleus to be imposed upon the electrons in the atoms. These new values will produce every time a different amount of amperes, and a different amount of developed resistance. The intensity of the temperature around the atom has no influence to make the electron consume more temperature than 1^0 Celsius. All this has been discussed in Chapter 17.

Another trick happening inside the atoms is that a small amount of thermodynamic temperature above zero Celsius makes the (e⁻) start moving around its (n). The electron needs one degree of temperature to produce its amperes and resistance per second with its (Ke). But what happens if the temperature around the atom is lower than 1^0 Celsius? In this case it has to borrow an amount of temperature by absorbing it from the space inside the atom and its surroundings! Doing this, the (Vol) and the (mDen) inside the atom will change and this is to be reflected upon the whole space of the universe! This is very important. When the temperature is increasing from 1^0 Celsius and up, the current (I) in amperes will be always decreasing, and the resistance increasing. But there is also (Vol) and (mDen) inside the atoms. Is there any change to them too? I believe there is. In higher temperatures the volume inside the atoms is increasing and the (mDen) decreasing. The higher (mDist) between the (n) and the (e⁻) will influence these factors too. As I have said before, **THERE IS A STIFF RELATION BETWEEN A TEMPERATURE INCREASE OR DECREASE IN THE NEIGHBOURHOOD OF THE ATOM, AND THE MAGNITUDE OF THE MEAN DISTANCE BETWEEN THE NUCLEUS AND THE ELECTRON** in meters. The changing (mDist) of any (e⁻) is very important to the circular electromechanics. I believe I **HAVE DISCOVERED THAT THE (mDist) IN METERS OF ANY (e⁻) FROM ITS NUCLEUS, IS ALWAYS THE SQUARE ROOT OF THE NUMBER OF DEGREES IN TEMPERATURE ABOVE ZERO CELSIUS IN THE VICINITY AROUND THE ATOMS, MULTIPLIED BY THE STANDARD** (mDist) OF THE (e-). For instance when the average thermodynamic temperature around the protium atom is 9^0 Celsius, the mean distance between the (e-) and the (n) is: $3 * (1 * 10^{-18})$. Below zero there is no motion of any (e-). They are stuck together with their nuclei, but my opinion is that (mDist) is calculated by assuming that the (e-) is still hypothetically moving around its (n), and in this case there is another raised problem. THE DEGREES OF TEMPERATURE BELOW ZERO CELSIUS I think might be EQUIVALENT TO THE SAME CORRESPONDING DEGREES WHICH ARE ABOVE ZERO CELSIUS WHEN THEY ARE RAISED TO THE 4th power. For example, the -2^0 Celsius could be equivalent to: $2^4 = 16^0$ Celsius above zero, and the -3^0 Celsius equivalent to: $3^4 = 81^0$ Celsius above 0^0. (See chapter 9). Let me talk now about the standard (mDist) between the nuclei and the electrons. Either the electrons are in the inner or outer cells, I think they get a standard (mDist) when the temperature around them is sharp 1^0 Celsius above zero. This (mDist) in the protium atom is $(1 * 10^{-18})$ meters. If an (e-) has a standard (mDist) or (mRd) between itself and the nucleus equal to $(2 * 10^{-18})$ meters, it means that this (mDist) has been fixed up when the temperature in the neighborhood of the atom was 1^0 Celsius.

Suppose that I have now the protium atom with a temperature around it equal to $.0001^0$ Celsius and then changed to 4^0 Celsius, and if I want to calculate their (FATT) at these temperatures first the (mDist) between the (e-) and the (n) of the protium atom at the temperature of $.0001^0$ Celsius, it is the square root of this temperature by the standard mean distance, which is: $\sqrt{.0001^0} * (1 * 10^{-18}) = 1 * 10^{-20}$, and its resistance between the (e-) and the (n) is $(1 * 10^{-20})^2 = 1 * 10^{-40}$. The (n) of this atom is the neutral proton which is neutrally charged and its (HP) is zero. The (e-) has an (hp) of $(-2/3 + 1/3) = -1/3$ but only when the temperature around the atom is 1^0 Celsius, so that the difference now in heat power between the positively charged (n) and the negatively charged (e-) at this changed temperature of $.00010$ Celsius will be different and equal to:

(HP-hp) = $0-(.0001 * -2/3 + .0001 * 1/3) = 3.3333333333 * 10^{-5}$. The (n) power will be: (HP-hp) $* c^2 = 3.3333333333 * 10^{-5} * 300000000^2 = 3 *10^{12}$. The (FATT) will be:

(HP-hp) $* c^2 / $ (mDist)$^2 = (3 * 10^{12}) / (1 * 10^{-40}) = 3 * 10^{52}$. The same protium atom at 4^0 Celsius around it will have a (mDist) equal to: $\sqrt{4} * (1 * 10^{-18}) = 2 * 10^{-18}$, and its resistance will be equal to: $(2 * 10^{-18})^2 = 4 * 10^{-36}$. The (HP-hp) between the (n) and the (e-) will be: $0-(4 * -2/3 + 4 * 1/3) = 1.3333333333$. This is a much higher difference between the (n) and (e-) in temperature than it was before when the temperature around this atom was $.0001^0$ Celsius.

The (n) heat power will be: $1.3333333333 * 300000000^2 = 1.2 * 10^{17}$, and the (FATT) will be equal to: (HP) of the nucleus minus (hp) of the electron multiplied by c^2 and divided by (mDist)^2.

It will be (HP-hp) $* c^2 / $ (mDist)$^2 = 1.2 * 10^{17} / (4 * 10^{-36}) = 3 * 10^{52}$. The same

(FATT) but the number of amperes and the number of resistance has changed. The (e⁻) in the atom with a temperature of .0001⁰ Celsius around it will have a higher speed in revolutions and it will cross more electric charges in one second and produce more amperes per second. The same atom with 4⁰ Celsius around it with a higher resistance between the (n) and (e⁻) will have less revolutions per second and will produce a lower number of amperes per second because it will cross less electric charges. If I divide the resistance of the protium atom at 4⁰ Celsius around it by the resistance of the same protium atom at .0001⁰ Celsius I will have $(4 * 10^{-36}) / (1 * 10^{-40})$ = 40000 lower resistance because of a lower mean distance in meters between the (e⁻) and the (n) and consequently increased production in amperes. In higher temperatures around the atoms the nuclear force becomes lower because there are lower amperes produced per second. Now I will be discussing the changes in volumes, mean densities, amperes and resistance in the protium atom and how they are calculated when average temperature around is changing from 1⁰ Celsius and higher and from 1⁰ Celsius and lower. Like with all other problems in this intractable universe, I think I have been able again to find out the right solution about this too. The classic equation in physics of

I = E / R as I have mentioned it before, in space is read like this: (FATT) or

(I) = (HP-hp) * c^2 / (mDist)². In space (I) represents the current just as it does in physics, (E) represents (HP-hp), and (R) represents (mDist)². The temperature in the neighborhood of the atom will regulate the (mDist) between the (n) and the (e⁻) in meters, the (n) power, the resistance of the (e⁻), and the amperes produced by its (Ke) in Kilogram-meters.

We are going to observe many reactions inside the atoms now, when the temperature changes to 1⁰ Celsius above zero and higher, or 1⁰ Celsius and down to zero. Starting from 1⁰ and higher, if the temperature goes up a corresponding increase of the (mDist) between the (n) and the (e⁻) in meters will occur, and almost everything will change, but more interesting now is the volume (Vol) and the mean density (mDen). The (Vol) inside the atom will increase in cubic meters to the third power of the changing (mDist) in meters, while the (mDen) simultaneously will get lower proportionally to the inverse of the volume increase to the 3ʳᵈ power. All these changes in volumes and mean densities inside the atoms simultaneously would be reflected on the whole universe. Let me suppose now that the temperature around the protium atom is 1⁰Celsius. At this temperature the mean radius of the (e⁻) is (1 * 10⁻¹⁸) meters. In this temperature no expansion or contraction in space could be possible. All kinetic energy in Kg meters produced and expressed in amperes and resistance would be spent just to balance the power of the nucleus. This electron in the protium atom of the hydrogen isotope with a (mDist) of (1 * 10⁻¹⁸) meters and in 1 degree temperature above zero will be the only one in the universe to produce 1 ampere and 1 ohm resistance in one second. In higher or lower temperatures around what ever is happening inside the atoms will be reflected in the whole space of the universe. Activated at 1^0 average temperature this electron would be forced to turn around its nucleus for one second and complete billions of revolutions in a millionth of a second. The energy of the thermodynamic temperature of 1^0 Celsius is the maximum that could be consumed by the (Ke) of the (e⁻) in 1 second. This 1^0 Celsius would be exactly what this (e⁻) in the protium atom would need to spin around its nucleus with the right number of revolutions per second, and therefore it could perform its duty integrally by producing the current of 1 ampere, and the resistance of 1 ohm in every one second. It is a total maximum performance that is to be exercised by this electron in the protium atom at this temperature of 1^0 Celsius. At 1^0 Celsius average temperature in space around the atoms, every electron has a standard (mDist) from its nucleus. I am sure and I want to make it clear that IN THIS THERMODYNAMIC TEMPERATURE, ALL THE MOVING ELECTRONS WILL HAVE NO EXTRA OR SPARE POWER TO EXPAND OR CONTRACT THE SPACE AROUND THEM. All their kinetic energy in Kg meters, expressed in amperes and resistance, would be spent just to balance the power of the nuclei as I said it above. Let us be aware, that if the temperature goes up and the (mDist) between the (e⁻) and the nucleus in any atom could remain the same, then its resistance would also be the same. The (e⁻) should sustain a higher (FATT) supplied by the nucleus, and by using its (Ke) it would have to be able to endure a higher pressure and temperature in its atom. For the (e⁻), it would mean that it should have to use the extra thermodynamic power of the increased degrees in temperature and by doing it, the (e⁻) also would have to increase the number of its revolutions for every second so that a greater number of its electric charges would have to be crossed over, and more power in amperes to be produced in order to balance the extra higher power of the nucleus. This thing could never happen in my opinion because the (e⁻) would never be able to consume more than 1^0 Celsius. That is the maximum the (e⁻) could do. That thermodynamic energy of 1^0 Celsius is all the power the (e⁻) would be able to return back to the (n) with its (Ke) transformed to amperes and resistance at the end of one second. It is as simple as that! The (e⁻) cannot afford to do any more. If it could, then it would be exceeding its maximum performance by taking the total BRUNT of the thermodynamic power of a total temperature, sometimes reaching billions of degrees. But the more INTERESTING point in this case would be that if the (e⁻) could remain with the same (mDist) from its (n), the (Vol) inside the atoms, and also the (Vol) of the space in the whole universe would not expand or contract. That would be an unprecedented event of cosmological dimensions. For example, in 1600^0 Celsius higher temperature, the power of the nucleus would be equal to: $0 - (1600 * -2/3 + 1600 * 1/3) * 300000000^2 = 4.8 * 10^{19}$. The resistance would be assumed to be the same as that of the electron in the protium atom at 1^0 Celsius. That should be: $(1 * 10^{-18})^2 = 1 * 10^{-36}$. And then the (FATT) would be:

$4.8 * 10^{19} / (1 * 10^{-36}) = 4.8 * 10^{55}$. This created (FATT) should be 1600 times higher than the standard (FATT) of $(3 * 10^{52})$, and this would have to be sustained by the (e⁻). You may see the difference if you divide the higher by the lower force of attraction. It would be: $(4.8 * 10^{55}) / (3 * 10^{52}) = 1600$. I do not think this could ever happen. Instead, I am rather quite sure that this amount of 1600^0 Celsius would increase the (mDist) between the (n) and the (e⁻) in meters, as well as the (Vol) in cubic meters inside the atom, and this all will be reflected in space, and thus expand the space of the whole universe. THE 13/27 (HP) RATIO OF THE PRE-EXISTING MASS WILL EXPAND THE MASS THROUGH THE ATOMS, AND THE ATOMS WILL EXPAND THE WHOLE UNIVERSE. The electron's (Ke) would use only 1^0 Celsius in one second to produce amperes and resistance, which multiplied together would equal the same power of 1^0 Celsius again. And the same (FATT) of $3 * 10^{52}$, multiplied by the resistance of this electron would equal the power of the nucleus. The resistance of the (e⁻) at the temperature of 1600^0 Celsius should be:$(4 * 10^{-17})^2 = 1.6 * 10^{-33}$. And if you divide it by the resistance of the electron in the protium atom at 1^0 Celsius temperature of $(1 * 10^{-36})$, you get 1600 times more. The (FATT) would show 1600 parts of resistance and 1/1600 amperes, $1600 * 1/1600 = 1^0$ Celsius. The protium atom as it is explained above has increased its (Vol) from the (Vol-1) at 1^0 Celsius to the (Vol-2) at 1600^0 Celsius by the amount of $((mDist2) / (mDist1))^3 * (Vol-1)$ or 40^3 multiplied by the (Vol-1).

The (Vol-1) should be $4\pi R^3 / 3$, and that would make:

$4\pi * (1 * 10^{-18})^3 / 3 = 4.18879020479 * 10^{-54}$ cubic meters. And then (Vol-2) would be equal to: $40^3 * 4.18879020479 * 10^{-54} = 2.68082573107 * 10^{-49}$ cubic meters. If you divide the (Vol-2) by the (Vol-1), you will get 64000. This number will indicate how many times more the (Vol-2) in cubic meters of the atom activated at 1600^0 Celsius has been increased from the (Vol-1) of the atom activated at 1^0 Celsius. The (mDen) inside the space of the atom activated at 1600^0 Celsius should be 64000 times less than that in the protium atom which was activated at 1^0 Celsius or:

$1 / 64000 * (2.68082573107 * 10^{-49}) = 4.1887902048 * 10^{-54}$. Inside the atom the percentage of the volume and density is: (Vol) = 1/(mDen) and (mDen) * (Vol) = 1.

$(4.1887902048 * 10^{-54}) / (2.680825173107 * 10^{-49}) = 1.5625 * 10^{-5}$ and $64000 * 1.5625 * 10^{-5} = 1$. Volume and mean density multiplied by each other will equal 1. All that has been said above concerning temperature, amperes, and resistance could be briefed up as follows: the number of degrees showing the new changing temperatures around the atoms every time would be the same numbers indicating the increased or decreased (HP-hp), the increased or decreased power of the nucleus, and the increased or decreased resistance in ohms and current in amperes for these atoms compared to the protium atom activated at 1^0 Celsius. These changes would not alter the number of the (FATT). It still would look the same as $3 * 10^{52}$. This is because the resistance of the (e⁻), with which the nucleus power is always divided, takes the same higher or lower value as the nucleus does. The same

$3 * 10^{52}$ is produced but with a different amount in amperes and resistance. Another thing I believe I found in this protium atom, is that at this magic temperature of 1^0 above zero Celsius, the time and the distance of the (e⁻) in meters per second go together in each one revolution, or each one completed cycle. It is $(2\pi * 10^{-18})$ meters in one cycle divided by $(2\pi * 10^{-18})$ time in seconds, will equal 1. Also, $(2\pi)^2$ electric charges will be completed in $2\pi * 10^{-18}$ meters or $2\pi * 10^{-18}$ seconds. In meters because, if I multiply one cycle of

$2\pi * 10^{-18}$ in meters by the number of revolutions made by the electron in 1 second, I will get the distance of 1 meter. And in seconds, because if I multiply one cycle of $2\pi * 10^{-18}$ in seconds by the number of revolutions completed by the electron to travel 1 meter distance, I will get the time of 1 second.

It is: $(2\pi * 10^{-18})$ in meters multiplied by $1.59154943092 * 10^{17} = 1$ meter.

And $2\pi * 10^{-18}$ in seconds multiplied by $1.59154943092 * 10^{17} = 1$ second. That looks like a wonderful performance made by this electron in the protium atom with the combined numbers of TIME, SPEED, and DISTANCE at 1^0 Celsius. The electrons in their atoms will expand the universe from the top high temperature down to 1^0 Celsius, up to where expansions inside the atoms will STOP SHARPLY. In general, the electrons are active from the highest point of temperature generated in our universe, up to the zero Celsius. The strangest thing will occur when the electrons will start to contract the space inside their atoms from 1^0 to 0^0 Celsius. For one only degree down the road the (e⁻) will nullify the total space between the (n) and (e⁻), if and only the whole universe has an average temperature of 1^0 Celsius to 0^0 Celsius. I was talking above about the reactions inside the atoms, which were taking place from 1^0 Celsius and higher.

What situations are created from 1^0 Celsius down to 0^0 Celsius, and then at 0^0 Celsius to the absolute zero degree? What volumes, mean densities, amperes and resistances are developed? In the classic electrical equation of $I = E / R$, the smaller the (R) is, the higher the number in amperes produced and vice versa. A similar situation is developed by an (e⁻) turning around its (n). Let me explain now what would happen to the protium atom activated in a temperature from 1^0 Celsius to 0^0 Celsius. In lower temperatures from 1^0 to 0^0 Celsius the (Vol) inside the atom will be decreased and the (mDen) increased. The amount of amperes and the strong nuclear force should be increasing rapidly. The number of amperes should be reciprocal again to the number of the degrees of temperature around the atom every second. To show you how rapidly these reactions are being done, let me suppose that the (e⁻)

in the protium atom is activated at .0001⁰ Celsius. At this temperature the amperes produced by the (e⁻) are 1/.0001 = 10000 and the resistance in ohms is 1/10000 per second. Indeed I think I can prove it.

First I take the (mDist) of the (e⁻) from its (n) which should be:

$\sqrt{.0001} * (1 * 10^{-18}) = (1 * 10^{-20})$ meters. Then I take the travelled (Dist) by the (e⁻) for one cycle, which would be: $(1 * 10^{-20} * 2\pi)$. And finally the revolutions per second would be equal to: $1/(2\pi * 1 * 10^{-20}) = 1.59154943092 * 10^{19}$. Dividing them by the revolutions per second of the (e⁻) in the protium atom of $1.59154943092 * 10^{17}$, I would get 100 times more. The crossed electric charges per second by the (e⁻) here should be: $1.59154943092 * 10^{19} * (2\pi)^2 = 6.28318530718 * 10^{20}$ times more. The $(2\pi)^2$ are the times of the crossed electric charges in one revolution by the (e⁻). If I divide the above number of the crossed electric charges per second by $(2\pi * 10^{18})$, which in one second produces 1 ampere, I would get 100 times more amperes than those produced by the (e⁻) in the protium atom activated at 1⁰ Celsius. The 100 times more electric charges now, multiplied by the 100 times more revolutions per second will make this (e⁻), which is activated at .0001⁰ Celsius produce 10000 times more amperes than it would produce in the protium atom activated at 1⁰ Celsius and 10000 times a stronger nuclear force. The resistance in ohms of course should be 1/10000. The thermodynamic temperature of 1⁰ Celsius is converted to a (Ke) by the (e⁻) to render 10000 amperes and 1/10000 resistance per second, and amperes multiplied by resistance will equal the nucleus power. The nucleus power is: $0-(.0001 * -2/3 + .0001 * 1/3) * 300000000^2 = 3 * 10^{12}$. And the (FATT) is: $(3 * 10^{12}) / (1 * 10^{-20})^2 = 3 * 10^{52}$. Same (FATT) but because the (mDist) between the (n) and the (e⁻) in meters has been decreased, the (Ke) of the (e⁻) will get higher, and there would be more revolutions per second to produce a higher number of amperes.

The (e⁻) consuming 1⁰ Celsius temperature again in one second, would bring down the temperature inside the atom, decreasing the (Vol) of space and increasing its (mDen). The decreased (mDist) or (mRd) will make the electron produce higher number of revolutions and more electric charges to be crossed by the (e⁻), and more amperes per second to be produced. The (FATT) imposed on the electron remains the same. In this situation the higher number of amperes produced will create and DEVELOP an extremely STRONG NUCLEAR FORCE between the electron and the nucleus. This is in my opinion what is happening in the atoms concerning the strong nuclear force. THE NUCLEAR FORCE IS ALWAYS RELATED TO A HIGHER OR A LOWER NUMBER OF AMPERES PRODUCED per second by the electron. The (mDist) of the (e⁻) from the nucleus will indicate how many amperes exactly will be produced per second, or how low or high the strong nuclear force will be.

The electron in the protium atom activated at 1⁰ Celsius will be the model to compare all other electrons how they are performing in the different situations that are being engaged. When the (e⁻) is turning around the nucleus from 1^0 to 0^0 Celsius it still consumes 1^0 Celsius per second, but the average temperature around the atom now is below 1^0 Celsius. The (e⁻) in this case I think has to borrow an amount of temperature from inside the atom and its surroundings, and in my opinion it will succeed to do so at the expense of the contraction of the volume inside the atom and in the surrounding space, but again this all could be done by the (e⁻) only if it could be in a position to develop a greater number of revolutions per second and a higher number in amperes per second. That of course could be easily obtained as it has been explained above. As the mean distance between the (n) and (e⁻) goes down approaching the zero point, so the number in amperes produced by the electron goes higher, and the strong nuclear force gets stronger.

The space of the atom inside could be easily reduced to zero, and the amperes at this point would get an infinite value, which would look like this, $(I = E / R)$ and if $(R = 0)$ then $(I = \infty)$. This whole thing is being done too fast, especially in the outer regions of the global universe where all of a sudden the electrons could contract the total space inside their atoms too fast, and whole galaxies could be disappeared and sucked by the "massive dark matter" named after by recent leading theorists. Let me now compare the volume of the protium atom starting from 1^0 Celsius and going down to $.0001^0$ Celsius. If this new (Vol) is named (Vol-2), and that of the protium atom at 1^0 Celsius (Vol-1), the (Vol-2) at $.0001^0$ Celsius is:

1 / ((mDist of Vol-1) / (mDist of Vol-2))³ * (Vol-1).

It is: $1 / ((1 * 10^{-18}) / (1 * 10^{-20}))^3 * (4.18879020479 * 10^{-54}) = 4.18879020479 * 10^{-60}$.

The $1/(mDist1/mDist2)^3$ is equal to: $1 * 10^{-6}$. And that if multiplied by the (Vol-1) will indicate the (Vol-2) in cubic meters and how many times smaller it would be than the (Vol-1). The (Vol-1) is: $4.18879020479 * 10^{-54}$ cubic meters. Multiply it by $(1 * 10^{-6})$, or 1/1000000 and you get: $4.18879020479 * 10^{-60}$ cubic meters. You may remark that the (Vol-2) has been 1000000 times less than the (Vol-1). To find out and verify it, divide the (Vol-1) which is $4.18879020479 * 10^{-54}$ by the (Vol-2) which is $4.18879020479 * 10^{-60}$. That will equal 1,000,000. You can imagine now how many atoms would be in a universe as it is observed today, when there are billions of atoms in a spot that could only be seen by an ordinary microscope. Let me now examine the reactions of the (e⁻) at 0^0 Celsius and below it. At 0^0 Celsius the resistance is zero and the amperes will become infinite. See Figure No.17 on the next page. There is no movement of the electrons here, and no chance of expansion or contraction. The electrons are consolidated with the nuclei. At this point and down below the stationary cold-dynamic powers start to work.

Fig No. 17

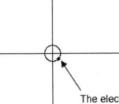

The protium atom is at a temperature of zero degrees Celsius. The electron is not moving around the (n). It has been short circuited with its (n). The mean distance between the (n) and the electron is zero. The current in amperes per second produced or the force of attraction is equal to the $(HP-hp)*c^2/(mDist)^2 = A/0 = \infty$. It is similar to $I = E/R$ when $R=0$, then the $I=\infty$.

The electron and the (n) together

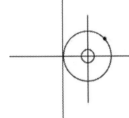

Here the protium atom is activated at 1 degree Celsius above zero. The electron has a standard (mDist) equal to $(1*10^{-18})$. The volume of the atom would never be expanded or contracted at this temperature. The current (I) in amperes produced here in one second is 1 ampere and the resistance 1 ohm.

The electron

The Nucleus (n)

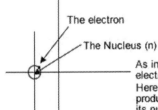

As in the first example, the electron is not moving around the (n). Here I believe, the amperes that would be produced if the electron was circling around its nucleus would be -1 ampere.

Fig No. 18

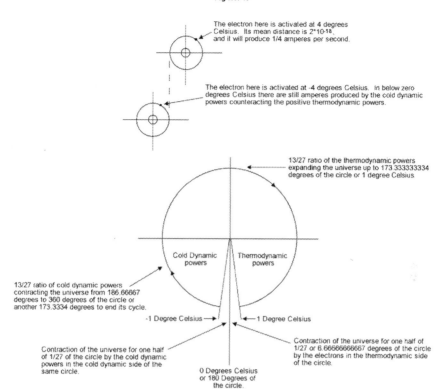

The electron here is activated at 4 degrees Celsius. Its mean distance is 2*10-18, and it will produce 1/4 amperes per second.

The electron here is activated at -4 degrees Celsius. In below zero degrees Celsius there are still amperes produced by the cold dynamic powers counteracting the positive thermodynamic powers.

13/27 ratio of the thermodynamic powers expanding the universe up to 173.333333334 degrees of the circle or 1 degree Celsius

Cold Dynamic powers

Thermodynamic powers

13/27 ratio of cold dynamic powers contracting the universe from 186.66667 degrees to 360 degrees of the circle or another 173.3334 degrees to end its cycle.

-1 Degree Celsius →   ← 1 Degree Celsius

Contraction of the universe for one half of 1/27 by the cold dynamic powers in the cold dynamic side of the same circle.

Contraction of the universe for one half of 1/27 or 6.66666666667 degrees of the circle by the electrons in the thermodynamic side of the circle.

0 Degrees Celsius or 180 Degrees of the circle.

All these interactions inside the atoms by the electrons is the **PHILOSOPHY OF OUR COSMOLOGY**. From -1^0 Celsius and below the (Vol) of the universe would start to decrease and the mean density to increase steadily to a higher rate in every second until the end of its cycle. At this point of my beyond standard model theory I was thinking about the energy and the period of time required for the total expansion of the universe that started from $5 * 10^9$ degrees of Celsius up to 1^0 Celsius, which might be the same and equal to that of the contraction. The energy and period of time required for the expansion is the same that is required for the contraction, which begins from 1^0 to

-273.16^0 Celsius, and this seemed like a short distance. But under 0^0 Celsius the numbers had to be raised to the fourth power in order to be equivalent to the external degrees of temperature. I thought a **STIFF RESISTANCE** would slow down the procedure to contract the universe taking a long time to finish, and probably as I said before to be equal to the time required for the total expansion of the universe.

Probably the slowing down of producing anti-amperes under zero temperature is the main reason I believe the universe could require a long time to finish those -273.16^0 Celsius and end its cycle. See Figure No.18. The expansion period of cosmic time for the electrons to move around their nuclei starts at $5 * 10^9$ degrees of temperature and ends at 0^0 Celsius. The mean distance of the electrons in the atoms (between themselves and their nuclei) in the temperature of $5 * 10^9$ is $\sqrt{(5 * 10^9) * (1 * 10^{-18})}$ = 7.07106781187 * 10^{-14} meters. Dividing it by the same standard mean distance of the (e^-) in the protium atom at 1^0 Celsius, I will have 70710.6781187 times greater (mDist) in meters of this electron. If I named the volume of the protium atom activated at 1^0 Celsius (Vol-1) and at $5 * 10^9$ Celsius (Vol-2), the (Vol-2) would be equal to:

70710.6781187^3 * (Vol-1) = (3.53553390594 * 10^{14}) * (4.18879020479 * 10^{-54}) = 1.48096097939 * 10^{-39} cubic meters larger. The universe is a container of cold and hot temperature. In my opinion the laws of Boilet, Charles, and Avogadro's, might be useful to our cosmology. I was thinking what would happen if the ratios of 13/27 and 14/27 hot and cold power was different in the pre-existing mass, and/or there was also a different top high temperature from $5 * 10^9$ degrees in the big bang as it has been suggested by the scientists, and verified by myself. Even a slightest difference I think would probably rapidly change our cosmology. This could be an unfortunate event. I don't think a universe could live without leptons and without planetary systems. I think that there must be a relation between the ratios of 13/27 and 14/27 hot and cold temperature existing in the universe and the total amount of the pre-existing mass in order to get a defined maximum volume of our finite universe. Also I believe what the physicists say about the first and the second law of the thermodynamics. They say energy is neither created nor destroyed, and also it is neither increased nor decreased. Then, the amount of energy in the whole universe should always be the same, never increasing or decreasing. (See pages 346-348 in the Encyclopedia Americana, 1997 edition). The dualism of the temperature in one entity is the belief in two antagonistic supernatural beings, the one good the other evil. Without leptons and atoms, there could be a deformation of our universe and a misfortune for our humanity. It seems to me that our cosmogony or the origin of creation of the world or universe, and/or the doctrine of the origin or formation of the universe, starts from one lepton that would need another two leptons to get together, and after that, another three leptons, and then another three, until there were 9 trios altogether to be joined by necessity, and then a microscopic world of a micro-cosmic universe could be created in the eons of time. This may be our cosmology, or the theory relating to it; it looks like that our infinite and finite universe belongs to these infinitesimal leptons consisted of hot and cold temperature, and this is our misleading world of nature and "THE ULTIMATE BEING OF OUR EXISTENCE". When the end comes a transformation of an abyss and Hades of a metaphysical world and a lifeless pre-existing mass would be formed. Another big bang should be on stand by! THIS ABYSS AND HADES IS "THE MASSIVE DARK MATTER" THAT HAS BEEN DISCOVERED BY OTHER THEORISTS IN THE OUTERMOST BORDERS!

CHAPTER 19

THE MASS CONVERTED TO ENERGY OF ELECTRO-VOLTS

WHY THE MASS OF AN ELECTRON IS EQUIVALENT TO 0.51 MEV'S?

I was very much and seriously concerned about this very important problem of converting mass into energy. It is a cosmological matter of the highest interest to our humanity. This I believe is still a puzzling problem for our hard working particle physicists. I believe I have the solution in my hands. I solved this problem like so many other ones, even though it seemed to be beyond my capabilities. I worked very hard indeed in order to find out the right path for the solution of this complex case in physics. Hopefully the solution for this problem will be a step forward to success and it will satisfy our particle physicists and scientists, and our mankind in general.

In my opinion their wonders, including that of Christine Sutton and Joseph Silk, editors of the books "BUILDING THE UNIVERSE" and "THE BIG BANG" should be over now. The scientists admit that the mass of an electron is about 9.1 * 10^{-28} gr., or using the scientific number it is: 9.1093897 * 10^{-31} Kg. This, they say is equivalent to: 0.51 MeV. I do not absolutely agree that each electron or positron has 0.51 MeV. Why is the electron's mass 0.51 MeV and not 0.5 MeV for instance as Christine Sutton wondered about? Referring to my cosmic theory in chapter (6), let me recall that the electron is an infinitesimal piece of mass of condensed temperature consisted of 2/3 cold-dynamic power which is to remain in that state in eternity, and 1/3 of thermodynamic power which is to expand the universe during a complete cycle. The thermodynamic mass of 1/3 of the electron then, must be the only amount of mass out of the total electron (e⁻), which would be converted into an equivalent energy of electro-volts. Similarly in the case of the positron, it is consisted of 2/3 thermodynamic power and 1/3 cold-dynamic power; the latter again is to remain in this state with no expansion or contraction forever in the eons of time. The amount of mass out of the whole infinitesimal mass of the positron, which is to be converted in to equivalent energy of electro-volts, is the 2/3 of it. This is an amount of thermodynamic power. Finally, the third elementary particle of neutrino has 50% heat and 50% cold powers. The heat power of the neutrino is 50% of the whole amount of its mass contained in itself, and this will be converted into an equivalent energy of electro-volts, that is 1 and 1/2 thirds out of its total of 3 thirds of mass. When the particle physicists achieved the head on collision of an electron (e⁻) and a positron (e⁺), they had obtained a rest mass energy measured in electro-volts of about over one million. Pay attention to this now, according to their estimation this amount of energy was for both electron and positron together.

It was supposed that each of these two particles of (e⁻) and (e⁺) had an equivalent amount of about 0.51 MeV, but this was not so! Their delicate and accurate apparatus showed the amount of the total equivalent energy of electro-volts, for both particles of the electron (e⁻) and positron (e⁺) together, but not for each one separately. The amount of energy for each one of the particles of the electron (e⁻) and positron (e⁺) would not necessarily be 0.51 MeV's. Of course the total amount found of electro-volts for both of them, when totally annihilated was over one million, but each one of the particles had a different equivalent amount of electro-volts! It was natural for the particle physicists to estimate that each one of these two particles should have an equivalent amount of rest mass energy equal to half of the total produced by both of them, (of over one million electro-volts) because they were not concerned about the construction of these two particles of (e⁻) and (e⁺) as well as the (v). In my cosmic theory I have made it clear (see chapter 8) that only the ratio of 13/27 heat power of the pre-existing mass will be converted to energy, while the other 14/27 ratio of cold power shall remain in the same condition not being expanded or contracted in eternity. This latter ratio is a stationary dynamic power, which will be used later to absorb the released energy from the 13/27 heat power ratio. Thus taking the case of the electron, it will have 1/3 of its mass converted into equivalent electro-volts, the positron will have 2/3 of its mass converted, and the neutrino 1 and a half thirds (or half of its total infinitesimal mass) of its mass converted into electro-volts.

From there on, I was thinking and thinking again, why it could not be feasible to solve the problem of how many electro-volts were produced exactly from each one of the electron, and/or the positron when they were totally annihilated during their head on collision. Finally, I came to the conclusion that the corner stone was the playing part of the three thirds of each lepton with different amount of heat and cold power in them. I inferred that the total rest mass energy for the two (e⁻ and e⁺), should be divided by three, (the 1/3 for the electron and 2/3 for the positron which makes 3/3 all together), and that was it! I found out that three times the 340,000 electro-volts would make it exactly 1,020,000 and this total amount divided by two would give 0.51 MeV. And conversely, dividing (0.51 MeV * 2) by 3, that is: (1,020,000/3) would give exactly 340,000 electro-volts for one fractional third. It was fantastic! Accordingly one electron and one positron together will return 1,020,000 electro-volts, 340,000 electro-volts for the (e⁻) and 680,000 electro-volts for the positron (e⁺), when they will collide head on, and both are completely annihilated! I would not hesitate to mention here that the rest mass energy of all kinds of leptons could also be verified as their total energy potential between the points of the absolute zero degree of temperature and the top high degree of about

$(5 * 10^9)$ Celsius per lepton. That, in my opinion has been the same degree of temperature per lepton that has been generated during the big bang (see chapter 21). This event of converting the mass of the leptons into energy was another proof that I have followed the right path of physics for my "Cosmic Theory of Thermodynamics". I will discuss in more detail how I am able to convert any amount of mass to an equivalent amount of energy. The weakest energetic particle of leptons is the electron (e⁻) with 1/3 heat power in itself, next is the neutrino (v) with one and a half thirds of heat power in itself, and last is the positron (e⁺), which is the strongest one with 2/3 of heat power in itself. Consequently, each one of them will yield 340,000, 510,000 and 680,000 MeVs, and all together a total of 1,530,000 electro-volts, when they will have a total rest mass energy. If I divide by three again the total number of electro-volts above produced by three leptons, I will get a mean value of 510,000 MeVs for each one them. A further analysis to the atom, by totally annihilating it will offer more proof and satisfaction to those that have not been yet totally satisfied with the leptons mass conversion into energy in the way I have explained it!

Indeed the scientists succeeded to smash a proton with its anti-proton as they say in the laboratories to yield about one billion electro-volts. But was it only one billion electro-volts for the two of them? One proton (P) contains 1836 leptons altogether, which are made up of (306 e⁻), (306 e⁺) and (1224 v). Their total energy is equal to: (306*340,000) + (306*680,000) + (1224*510,000) = 936,360,000 electro-volts, close to 1,000,000,000 electro-volts. And that is for one proton only. The same results will be obtained if I divide the total fractional charges of thirds contained into the proton by two, that is: (1,836 * 3 = 5,508) divided by 2 will be 2,754, and multiplying this 2,754 by 340,000 it will be equal to: (2754*340,000 = 936,360,000). These 2,754 fractional thirds is the portion of the mass representing the thermodynamic power of the nucleus of the protium atom.

I want now to make things more exciting, going further to find out the rest mass energy in electro-volts of any amount of mass as I have done it above. I could take for instance the total mass of a proton equal to 1836 leptons, multiplied by the mass of a lepton in kilograms, which is equal to: $9.1093897 * 10^{-31}$ Kg. And that would amount to: $1.67248394892 * 10^{-27}$ Kg.

In 1 Kg then there would be: $1 / (1.67248394892 * 10^{-27})$ Kg = $5.97913062571 * 10^{26}$ protons. Or if it were for a neutron, I would have the mass of 1 lepton multiplied by 1839. That would make the mass of one neutron equal to: $1.67521676583 * 10^{-27}$ Kg.

Dividing 1 Kg now by the mass of 1839 leptons in 1 neutron, I will have the number of neutrons contained in 1 Kg. This will be: 1 / (1.67521676583 * 10^{-27}) = 5.96937674215 * 10^{26}(there are less neutrons than protons in the mass of 1 Kg as you may remark). If 1 kg of mass contains 50% protons and 50% neutrons, I will add the mass of one proton and the mass of one neutron in Kg, and I will divide them by 2. That is:

(1.67248394892*10^{-27} Kg + 1.67521676583 * 10^{-27} Kg) / 2 = 1.67385035738 * 10^{-27} Kg.

Finally, dividing 1 Kg by this number of 50% protons and 50% neutrons, I will have approximately: 1 / (1.67385035738 * 10^{-27}) = 1 Kg of 5.97424970273 * 10^{26} protons and neutrons together. Each one of them has an energy in electro-volts equal to 936,360,000. After that it will be very easy to calculate the number of electro-volts produced by 1 Kg of protons, 1 Kg of neutrons, or even 1 Kg of 50% of each one of them. If I would choose to find out the rest mass energy in electro-volts of 1 Kg of mass made out of protons, I should multiply 5.97913062571 * 10^{26} protons by 936,360,000 electro-volts and this will be equal to: 5.59861875269 * 10^{35} electro-volts. I recall at this point that 1 proton has 1,224 neutrinos, and 5.97913062571 * 10^{26} protons will contain 7.31845588587 * 10^{29} neutrinos. I will remind the physicists and astronomers about the supernova or neutrino stars in space. Why are they becoming so bright when they explode? Also another thing that probably might have to be looked at is the already universally accepted famous equation of Einstein ($E=mc^2$). I do not know if it has to be corrected when and after my theory is judged and possibly accepted. My opinion is that it has to be changed to a different form of expression. I think what I said about the leptons is correct. I have talked quite extensively about their accurate and delicate precision of construction by birth, and their relation to the ratios of 13/27 and 14/27 of the thermodynamics and cold-dynamics consecutively, and that these two ratios are condensed in the total pre-existing mass. This all I believe probably remains one of the biggest cosmological events in my theory. I dare to go further and say that in my opinion some things in classic physics might have to be changed. In my view an entirely new path has to be followed in this branch of science since the discovery of these famous leptons, and their thermo/cold-dynamic powers intrinsically condensed. We should not forget that my theory is beyond standard model. I think that previous theories and laws of physics may have to be looked at. As an example, as I already mentioned on this page, one of these laws is Einstein's equation ($E=mc^2$). We learned and now know the ratios of thermodynamic and cold-dynamic powers in the universe are 13/27, and 14/27. Thus in any amount of mass only the 13/27 of it will be converted into an equivalent amount of energy in electro-volts. Einstein's equation of $E=mc^2$ refers to the total mass I believe and if it was ever changed, it would look like: $E=(13/27)*mc^2$.

CHAPTER 20

CLARIFICATION OF THE REQUIRED NUMBER OF ELECTRO-VOLTS TO PRODUCE THE POWER OF ONE WATT IN COSMIC PHYSICS

AT THIS POINT again, I was thinking about the amount of the electro-volts required to produce one watt. In classic physics the particle physicists I believe have defined the unit of 1 electro-volt as being equal to the energy gained by a single electron, accelerated through a potential difference of 1 volt. (See page 105 from the book "Big Bang" by Joseph Silk, 1982 edition. It says that 1000 billion electro-volts equal 1 erg). Erg comes from the Greek word "ergon" which means work. On the Internet you may find the definition from WhatIs.com, which says that erg is a small unit of energy, equivalent to 0.0000001 of a joule. Energy in the centimeter gram second (cgs). On the Internet, at Answers.com, I found that an electro-volt is a unit of energy acquired by an electron falling through a potential difference of one volt, approximately $1.602 * 10^{-19}$ joules. Physicists I think have estimated that 10,000,000 ergs of energy approximately expended for one second, amount to a power of one watt. I paid my attention to the total number of electro-volts for one second. There I think they found about 10 million ergs, multiplied by 1,000 billion electro-volts should equal ($10 * 10^{18}$) electro-volts, and if expended for one second they would produce the power of one watt. But they did not have in mind how many electro-volts could be produced by the kinetic energy of an electron orbiting around its nucleus of the protium atom for one second in order to determine the accurate number of electro-volts for this important matter. To the best of my knowledge I have tried quite a few times to change some mathematical laws of the classic physics, just to fit my theory of cosmic physics, which is beyond standard model so that I could obtain good results and I hope that responsible historians and scientists will sometimes justify me and accept my theory. The $10 * 10^{18}$ electro-volts estimated by the physicists probably a long time ago, to produce one watt in one second was close to the number of electric charges, that one (e⁻) has to cross through in one second in the line of its elliptical orbit around the nucleus of the protium atom to produce the power of one watt. This number is exactly

($2\pi * 10^{18}$ and it has been explained in previous chapters) and in my view this is the number of electro-volts expended for one second to produce the power of one watt. I conclude then, that if there still is this number of $10 * 10^{18}$ electro-volts to produce one watt it might have to be changed to: $2\pi * 10^{18}$ electro-volts instead. I think without any doubt this is the right number.

CHAPTER 21

RELATION OF TEMPERATURE, VOLUME AND DENSITY- TOTAL AND/OR PARTIAL RELEASE OF ENERGY OF OUR UNIVERSE IN ELECTRO-VOLTS

THE TOTAL PRE-EXISTING mass of the universe is the two ratios of hot and cold powers (13/27) and (14/27) consecutively, and they would define the size of the whole universe. The ratio of 13/27 is the thermodynamic power ratio of the pre-existing mass and the only one portion out of the total mass, which will regulate the magnitude of the volume of space that will be expanded. The idle portion of the other ratio of 14/27 of the pre-existing mass is the cold-dynamic power, which will absorb and contract the expanded space of the universe when the time comes. All these could be demonstrated in a circle diagram, for a complete cycle of the universe. See figures 8, 9, 18, 19, and 20. Meanwhile, the total energy release in this universe will be an immense work lasting probably 4π billion years including both expansion and contraction, as I do believe myself. The 13/27 heat power ratio of the total pre-existing mass will bear the whole burden of the total space expanded in our universe. When all its energy of this ratio is released, the average temperature in the universe will stand at 1^0 Celsius. I believe that it is a spherical and finite universe inflated and deflated perpetually in the eons of time, and also I do believe that the lasting period of both expansion and contraction of our universe are equal. Now I think that the estimated top high absolute temperature of $5 * 10^9$ per lepton that was estimated by the physicists when the big bang occurred is the right one, and it always should be the same and never increasing of decreasing. A little later I will explain how I found out that probably this was and it will be forever the right top temperature in the universe indeed. When the average temperature in the universe reaches 1^0 Celsius, the maximum expansion will be shown in the cycle diagram and when the average temperature reaches 0^0 Celsius this would be the middle line where higher or lower temperatures will expand or will contract the universe. For the interval between 1^0 to 0^0 Celsius the remaining energy of the total intrinsic top high temperature inside the leptons would be consumed to contract all the space of the atoms and by doing it they will contract the whole universe drastically.

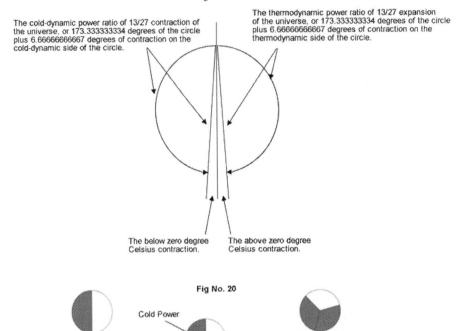

Fig No. 19

The cold-dynamic power ratio of 13/27 contraction of the universe, or 173.333333334 degrees of the circle plus 6.66666666667 degrees of contraction on the cold-dynamic side of the circle.

The thermodynamic power ratio of 13/27 expansion of the universe, or 173.333333334 degrees of the circle plus 6.66666666667 degrees of contraction on the thermodynamic side of the circle.

The below zero degree Celsius contraction.

The above zero degree Celsius contraction.

Fig No. 20

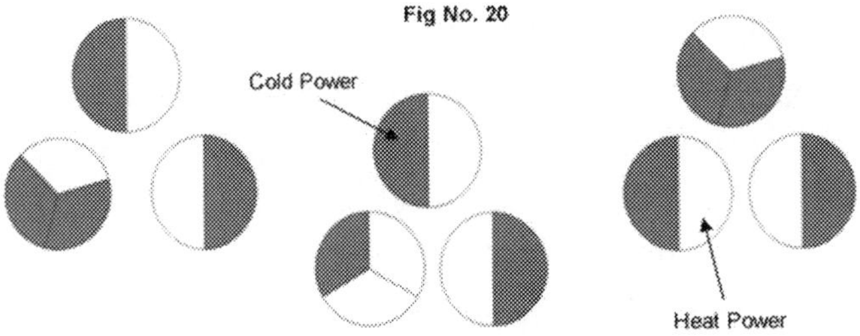

Cold Power

Heat Power

Here there are 14 thirds cold and 13 thirds hot power. These 9 leptons with 27 fractional charges of thirds constitute a micro-cosmic Universe. Each one of these leptons consists of three fractional thirds. They have a shape of a sphere and each one of these leptons is intrinsically filled up with thermodynamic and cold-dynamic powers.

The percentage of the volume (Vol) and the mean density (mDen) in the universe is always equal to 1/(Vol) = (mDen) and 1/(mDen) = (Vol), whether the universe is being in the expansion or in the contraction period. That is: (Vol) * (mDen) = 1, and (Vol) = 1/(mDen), and also (mDen) = 1/(Vol). The average absolute temperature in space starting from $5*10^9$ degrees of Celsius per lepton approximately as I believe, and ending up in the absolute zero degree, will indicate the different positions of our universe and the magnitude of its (Vol) and (mDen) in its circle diagram at any cosmic time. Figure No.19 on this page shows the number of degrees held in the circle diagram, by each one of the thermodynamic and cold-dynamic ratios in the whole universe. There are 173.333333334 degrees out of the 180 degrees in the circle, or a percentage of 0.481481481% for the expansion period plus 6.66666666667 degrees out of the total 180 degrees, or a percentage of .018518518519% for the contraction period. Figure No.20 shows the number of leptons, which could represent a micro-cosmic universe. There are 2 electrons, 1 positron and 6 neutrinos in it, or 1/3 of electrons and positrons together and 2/3 of neutrinos.

The whole universe is consisted of a multiple number of the above proportion. In general the universe as a whole is constructed by this same proportion of all these three kinds of leptons. It is a mystery of the nature, and it is also a symbolic representation of the "TRINITY" (see chapter 6). These 9 leptons as you may remark, are consisted of two ratios. One ratio is the 13/27 fractional charges of thirds of thermodynamic power, and the other ratio is the 14/27 fractional charges of thirds of the cold-dynamic power. That means that the 13 out of the total 27 fractional dynamic charges are to be the energetic ones, to expand the universe.

I will recall that the protium atom consists of 306 electrons, 306 positrons and 1224 neutrinos, multiplied by 3 will make up 5508 fractional charges of thirds altogether. I think that under scientific estimation, the smallest speck that can be seen by an ordinary microscope may contain millions of atoms. Also see page 633 of the Encyclopedia Americana that says there are 50 million atoms ranged side by side in one linear cm of solid matter. I repeat here that the total energy of the top high temperature of $5*10^9$ was totally consumed when the average temperature in space came down to 1^0 Celsius, EXCEPT for the intrinsic energy of the top high temperature inside the leptons themselves. This energy would be consumed totally and drastically from 1^0 to 0^0 Celsius, and then the cold-dynamic powers would take over to absorb the previously total released energy from the thermodynamic powers and contract the universe. But what unbelievable facts might be disclosed at 0^0 Celsius?

At 1^0 Celsius the 13/27 ratio of the pre-existing mass has completed the total and maximum expansion of the space in the universe. This is the energy of one kind of temperature of how hot or how cold something is in the open space in the universe (see chapters 6 and 8). The other kind of temperature is that inside the leptons and this is the top high and bottom low temperature intrinsically condensed in the leptons inside the atoms (see again chapter 6). The microscopic universe which represents the whole universe on the other hand is consisted of 9 leptons with 27 fractional thirds and only the 13 of them have the top high energetic thermodynamic power in themselves. This 13/27 ratio will be responsible to consume its total mass energy to expand the space of the universe from about $5 * 10^9$ degrees of Celsius per lepton up to 1^0 Celsius. When the temperature reaches an average of 1^0 Celsius in the whole space of the universe, there will be no more expansion anymore, although 1^0 Celsius average will still be remaining in the universe. This 1^0 Celsius is consumed from the still agitated electrons turning around the atoms and until the average temperature in the universe reaches 0^0 Celsius. From 1^0 to 0^0 Celsius all the rest mass energy of the total leptons in the atoms have been consumed. This 1^0 Celsius is concentrated intrinsically inside the leptons of the atoms, and all of the leptons in the universe compose the pre-existing mass with the two ratios of 13/27 heat power and

14/27 cold power. Now there are three different basic leptons that have made the universe. They are the electron, the positron and the neutrino. If I take two electrons, one positron and 6 neutrinos (see Figure No.20) I would have 9 leptons altogether and with these 9 leptons a microcosmic universe would be prepared with the same two ratios of 13/27 heat power and 14/27 cold power. This proportion is the same, as it exists in the whole universe. These two ratios of 13/27 and 14/27 have 13 fractional charges of thirds in heat power and the other 14 fractional charges of thirds in cold power; all of them contained inside the 9 leptons of the microcosmic universe. If I totally annihilate the mass of all of them,, I would have the rest mass energy in electro-volts, produced only by the heat power ratio of 13/27 and then if divided by the 9 leptons I would have the electro-volts produced per lepton during the big bang and if these electro-volts were transformed or converted to thermodynamic temperature I would know what the top high temperature was during the big bang. In chapter 19 I explained in detail how I have discovered the rest mass energy of all these three kinds of basic leptons, which I am positively sure they are the blocks of the nature that made the whole universe, and so these 9 leptons may be functioning the same exact way as our immense universe does.

The total number of electro-volts that would have been produced by these 9 leptons after totally being annihilated should be the energy of the 13 fractional thirds out of the total 27, which multiplied by 340,000 electro-volts will equal to 4,420,000 electro-volts. If these electro-volts are divided now by the 9 leptons, I will get 491,111.111 electro-volts for each one lepton, which is very close to the number of 500,000 electro-volts. These 500,000 electro-volts now, according to the scientific estimations will be equivalent to 5,000,000,000 degrees of Kelvin (see page 110 from the book "The Big Bang" The Creation And Evolution of the Universe by Joseph Silk for the conversion of electro-volts to Kelvin degrees). But this number of $5*10^9$ degrees in temperature is approximately the number of the absolute temperature in degrees of Celsius per lepton existing during the first picoseconds of the Big Bang! Could you imagine this concentrated heat power inside a lepton or inside an atom? Again, there are 50 million atoms set side by side in a linear distance of 1 cm of solid matter, and in each proton there are 1836 leptons, and in each neutron there are 1839 leptons. (See page 871 in the World Book Encyclopedia).

From all of this I deduce, that if I know some of the factors in our universe, I might be able to make it feasible, to calculate a partial or even the total energy released in electro-volts in the universe. Scientific estimation indicates that the amount of mass of one lepton is: $9.1093897*10^{-31}$ Kg, (see page 694 in the Encyclopedia Americana) and the number of leptons in one Kg of mass should be: $1/9.1093897*10^{-31} = 1.09776838288*10^{30}$ approximately. Each one Kg contains $5.97424970273 * 10^{26}$ protons and neutrons (see page 132). Each one proton or neutron has an energy in electro-volts equal to 936,360,000, so each one Kg of mass contains 936,360,000 * 5.97424970273 *10^{26} electro-volts approximately. For the whole universe the energy in electro-volts would be approximately equal to: the total pre-existing mass in Kg multiplied by 936,360,000 * 5.97424970273 * 10^{26}. And this is the total energy in electro-volts released in order to create this universe. I would get the same results if I multiply the total number of leptons by 5 * 10^9 Kelvin temperature and change this temperature in electro-volts.

CHAPTER 22

THE SOLAR SYSTEM AND OUR GALAXY OF THE MILKY WAY

THE SUN BY scientific information has been found to have 0.998 of the total mass in our solar system. I used the method of multiplying the volume by the density in order to find out the amount of mass in Kg of any celestial body. As you may see in chapter 15, the mass of the sun is found to be equal to: $3.67189011568*10^{26}$ kilograms. This number constitutes the 0.998 mass of the total mass in the solar system. The remaining mass of

1-0.998 = 0.002 belongs to the nine planets and their satellites. All the above said is according to scientific information. The total mass of the solar system should be approximately equal to the mass of the sun, plus the mass of the nine planets and their satellites. The mass of the planets and their satellites in our solar system is approximately equal to: $3.67189011568*10^{26} * 0.002 = 7.34378023136*10^{23}$ Kg. And the total mass of the solar system should be equal to about: $3.67189011568*10^{26}$ Kg + $7.34378023136*10^{23}$ Kg = $3.67923389591*10^{26}$ Kg. Taking the information from the encyclopedias about the "Milky Way" galaxy, I have learned that its mass is equal to about 100 billion solar masses, that is:

$(3.67923389591*10^{26}) * 10^{11} = 3.67923389591*10^{37}$. This is the mass of the Milky Way galaxy. Also, according to scientific information every 100 solar masses will make one unit of a solar luminosity for the Milky Way. The 100 solar masses are equal to: $100*3.67923389591*10^{26} = 3.67923389591*10^{28}$.

Dividing the total mass of the Milky Way, which is $3.67923389591*10^{37}$ by the 100 solar masses of $3.67923389591*10^{28}$, I will get 1,000,000,000 units of solar luminosities for our Milky Way galaxy. The same thing is obtained if the number of 100 billion solar masses is divided by the 100 solar masses. That is:

100,000,000,000/100 = 1,000,000,000. One solar luminosity actually is equal to the mean temperature per lepton in the mass of the sun, which has a temperature of about 7000000^0 Celsius. This is one unit of solar luminosities. The Milky Way has about 1,000,000,000 of these units approximately. Multiplying 7000000^0 Celsius by 1,000,000,000, I will have the number of total luminosities for the Milky Way. That is: $(7*10^6)*(10^9) = 7*10^{15}$. That is approximately the luminosity of the galaxy. Its heat power (HP)is the:

(tM) *$(7*10^{15})$ = $3.67923389591*10^{37}*(7*10^{15})$ = $2.57546372714*10^{53}$. The total power of that galaxy is approximately: (HP)*$(c)^2$ = $2.31791735442*10^{70}$. The mean distance or the mean radius of our solar system from the centre of the Milky Way estimated by the scientists is approximately 30,000 light years away, or three fifths from the center to the edge of the galaxy. In order to find this (mDist) or (mRd) in meters, I multiply the time in seconds of one light year by 30,000 years, and then this answer by the speed of light per second. It will be: (31558150*30,000)*300,000,000 = $2.8402335*10^{20}$ meters.

The traveling distance of the solar system of $(2\pi R)$ around the galaxy would be: $2.8402335*10^{20}*2\pi$ = $1.78457133962*10^{21}$ meters. The electric charge will be: $(1/2\pi)$*(mDist) or (0.159154943092)*$(2.8402335*10^{20})$ = $4.5203720106*10^{19}$. It is the resisting power in ohms, of the solar system to the force of attraction of the Milky Way, for a travelling distance of one only meter on the line of its elliptical orbit around the galaxy. The scientists have estimated that the solar system will complete one cycle around the centre of the Milky Way every 225 million years approximately, with a speed of about 250,000 meters per second. Multiplying this number of years by 31558150 (number of seconds for one year) I get: $7.10058375*10^{15}$ time in seconds for one cycle of the solar system around the galaxy. If I multiply this time in seconds by the speed of the solar system in meters per second, I will have its total travelling distance of $(2\pi R)$ in meters equal to: $1.7751459375*10^{21}$, and this divided by (2π) will give almost the same (mDist) in meters of $2.82523250663*10^{20}$ as it has been calculated above, where the (mDist) of the solar system was estimated to be 30000 light years away from the center of the galaxy.

It is useless to say that when I have this (mDist) and the speed of the solar system in meters per second, I may calculate all other factors. If I use the mean radius calculated previously (by converting the 30000 light years into meters) I will have the same number for the (mRd) of $2.8402335*10^{20}$ and travelling distance equal to: $1.78457133962*10^{21}$ in meters. Then dividing now this travelling distance by the speed of 250,000 meters per second, I get the time in seconds of: $7.13828535846*10^{15}$ needed for one revolution of the solar system around the galaxy. Almost the same as when it was calculated above by multiplying the 225,000,000 years by the time in seconds for one year. This will indicate that the number of 30,000 light years, the speed of the solar system in meters per second, and the time in millions of years for one revolution that were estimated by the scientists have almost been accurate. You may divide either the total travelling distance found above of the solar system in one revolution by the speed of 250,000 meters per second, and you will get: $1.78457133962*10^{21}$/250,000 = $7.13828535848*10^{15}$ seconds, and $1.7751459375*10^{21}$/250,000 = $7.10058375*10^{15}$ seconds. Divide these different times in seconds for one revolution, by the time in seconds for one earthly year, and you get 226194671 and 225,000,000 millions of years.

Using the equation of the circular electro-mechanics, the revolutions of the solar system for one meter are:

(el-ch) / (mDist)^2 = 4.5203720106*10^{19}/(2.8402335*10^{20})^2 = 5.60358657455*10^{-22},

or:

1 / (1.7751459375 * 10^{21}) = 5.63333965324 * 10^{-22}. The total resistance of the solar system for one complete revolution around the galaxy is the (mRd)^2 of (2.8402335*10^{20})^2 equal to: 8.06692633452*10^{40}. Or multiply the (el-ch) by the total travelling distance (2πR) of the solar system around the galaxy, which is: 4.52037201 06*10^{19}*1.78457133962*10^{21} = 8.06692633452*10^{40}.

When the centre core of the heat power of the galaxy is connected with other galaxies or solar systems like ours, it will develop potential differences between themselves. As a result, tremendous electromotive forces between the poles of them will be generated simultaneously, as it is happening with any other planetary system in space. These absorbing powers of potential differences transformed into watts, will develop the forces of attraction to the negatively charged masses of stars or galaxies.

The equation of the force of attraction is: (HP-hp)*c^2/(mDist)^2. The total heat power of the Milky Way is: 2.57546372714*10^{53}. The total heat power of the solar system is approximately: 3.67923389591*10^{26}*7,000,000 = 2.57546372714*10^{33}. All other factors are known. The force of attraction between them is:

(FATT) = (2.57546372714*10^{53}-2.57546372714*10^{33})*c^2/(mDist)^2.

That is: 2.57546372714*10^{53}*300,000,000^2/(mDist)^2, or: 2.31791735442*10^{70}/(2.8 402335*10^{20})^2 = 2.87335877173*10^{29}.

If the speed of light is about 300,000,000 meters per second, its travelled distance for one year will amount to: 31,558,150*300,000,000 = 9.467445*10^{15} meters. The 30,000 light years, multiplied by 9.467445*10^{15} will equal to: 2.8402335*10^{20} meters! It's the distance of the mean radius in meters between the poles of our solar system and the galaxy. And vise versa dividing the (mRd) by one light year in meters of 9.467445*10^{15}, you get back the 30,000 light years. The total travelling distance then, of (2πR), would correspond to: (1.78457133962 * 10^{21}) / (9.467445 * 10^{15}) = 188495.56 light years distance taken by the solar system for one revolution around the Milky Way.

1.78457133962*10^{21} is the travelling distance in meters of our solar system to complete almost a circular orbit around the centre of the Milky Way every 225 million years. Also, 30000*2π = 188495.56 light years (see Figure No.21). The electric charge of the solar system is the resisting power for one only meter travelling distance on the line of its elliptical orbit. It is a resisting power of the kinetic energy of the solar system moving for 1 meter equal to: 4.5203720106*10^{19} kilogram meters equivalent to coulombs, and multiplied by the total travelling distance of 2πR around the galaxy, it would amount to the total resistance of 8.06692633452*10^{40}.

This resistance is developed during one complete revolution around the Milky Way. The (FATT) of the Milky Way when it is connected with our solar system electrically will be $(HP-hp)*c^2/(mDist)^2$.

This will create a huge potential difference that is to be balanced in one complete revolution by the (Ke) in Kg meters from the solar system around the galaxy. This (Ke) could be expressed in negatively charged amperes and resistance produced by the solar system and multiplied together would be returned back to equal the total power of the galaxy in WATTS. Exactly similar to the classic electrical equation of $(E=I*R)$. See previous chapters.

The (FATT) or radiation in watts per second imposed to the solar system is equal to: $2.87335877173*10^{29}$. The total resistance developed by the solar system is equal to: $8.06692633452*10^{40}$. The total (Ke) of the solar system is the produced amperes multiplied by the resistance developed. These two multiplied together will equalize the power of the Milky Way imposed on the solar system, which is equal to $2.31791735442*10^{70}$. It is exactly the same performance shown by the earth and the moon around the sun, or the electron around the nucleus. If I wanted to find our how many amperes are produced by the solar system in one second, it would be as easy as when I was doing the same calculation on the earth-moon.

In one second the solar system will produce a useful work equal to the speed of the solar system per second divided by $2\pi R$ multiplied by the (FATT) or: $(250,000)$ / $(1.78457133962*10^{21})*2.87335877174*10^{29} = 4.02527865929*10^{13}$ amperes. I am just multiplying the travelling distance of the solar system in revolutions for one second by the total (FATT) in amperes per second, which the solar system will accomplish in one complete cycle. The resistance for one second will be found in the same way. I will multiply the same travelling distance in revolutions for one second by the total resistance, which would be developed by the solar system in one complete cycle. The travelling distance in revolutions for one second would be: $250,000$ / $(1.78457133962 * 10^{21}) = 1.40089664363 * 10^{-16}$. The total developed resistance in one complete cycle is equal to: $8.06692633452*10^{40}$. The resistance developed for one second should be equal to:

$1.40089664363*10^{-16}*8.06692633452*10^{40} = 1.13009300265*10^{25}$. The total work produced for one second of the solar system around the galaxy is: $I*(mDist)^2$ or $(4.02527865929*10^{13})$ amperes$*1.13009300265*10^{25}$ resistance $= 4.54893924658*10^{38}$. For the total seconds of $7.13828535846*10^{15}$, the solar system would need to travel one complete cycle. The total work, $I*(mDist)^2$ or $(FATT)*(mDist)^2$ would be equal to the time in seconds for one complete revolution in the square, multiplied by the work of the solar system accomplished in one second. That would be: $(7.13828535 846*10^{15})^2*4.54803924658*10^{38} = 2.31791735442*10^{70}$. The same power in watts per second of the MILKY WAY, which is $(HP-hp)*c^2$ imposed on the solar system. This is again the same procedure as that between the sun and the earth-moon explained in chapter 15.

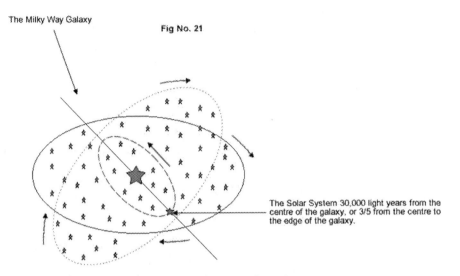

The Milky Way Galaxy

Fig No. 21

The Solar System 30,000 light years from the centre of the galaxy, or 3/5 from the centre to the edge of the galaxy.

The Solar System will travel around the centre of the galaxy (dashed line) with a speed of 250,000 meters per second, and it will take approximately 225,000,000 years to complete one revolution around the centre of the galaxy. This traveling distance (2πR) of the solar system is about 1.78457133962*10 ^{21}meters, or 188,495 light years.

CHAPTER 23

THE UNIVERSE IS ONE PLANETARY SYSTEM DISPUTED BY TWO SUPER POWERS OF DIFFERENT MAGNITUDE: THE THERMODYNAMICS AND THE COLD DYNAMICS

YES, THE UNIVERSE is one planetary system as a whole. Now as I am approaching the end of my cosmic theory which has been a cosmological mystery, I will go ahead with a synopsis of how our finite universe is functioning as a whole in short. To understand it thoroughly, let me suppose that the **whole universe represents a sole gigantic planetary system, of one local thermal equilibrium**. And in reality, this is true! Two huge super powers inconceivable in magnitude, govern this planetary system. (Figure No.22 will show more details). Namely these super powers are:

a)The 13/27 ratio of thermodynamic power over the total of heat and cold powers of the pre-existing mass in the universe.

b)The bigger ratio of 14/27 cold-dynamic power of the same pre-existing mass.

These two super powers exert titanic forces of mutual attraction to all the kinetic masses swinging around the space of the universe in two different cosmic periods of time. The one period starts from the very beginning of the Big Bang, and terminates when the average temperature in space drops down to zero degrees of Celsius. In that period the thermodynamic powers are prevailing. The second cosmic period starts from that point of zero Celsius degrees average in the whole space of the universe, and it finishes when it comes down to the end of an absolute zero degree. During this latter period the cold-dynamic powers are taking the upper hand. Hypothetically this thermodynamic power, which is the (HP) of the pre-existing mass is located in the center of the globular universe, and the cold-dynamic power which is the (CP) of the pre-existing mass is in the outermost layers of the sphere of the universe.

These two almighty powers exert tremendous forces of attraction to all negatively charged masses in space. They both try to pull the disputed masses in space on their side as long as the universe is alive. Both these super powers constitute the total of the **hot and cold masses** swinging around all over in space. They are being diametrically one opposite the other as Figure No.22 shows. When the universe is active the thermodynamic powers are winning these masses by pulling them around on their side, to create planetary systems. But this is not being done perfectly and at no expense without some reaction by the cold-dynamic powers in the universe. These latter ones are also pulling the kinetic masses to their side. The result is that kinetic masses turning around will change their direction to an elliptical or even a spiral one. At the same time centrifugal forces would be generated. (See Figure No.22).

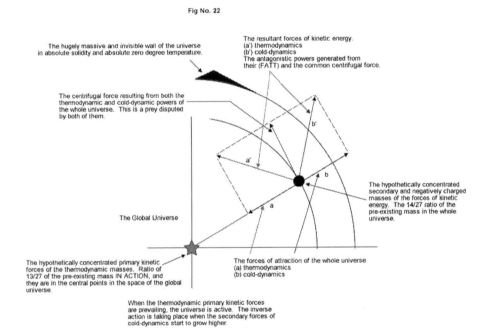

Fig No. 22

The cold-dynamic powers are always exploiting the growing volume of space, which gives them more power to react against the thermodynamics. As the volume gets larger, the cold-dynamics are taking momentum, especially during the end of the space expansion in the universe. The thermodynamic powers dominate the center of the sphere, while the cold-dynamic powers are always predominant at the extreme perimeters of the globular universe. At the beginning of the expansion the hot masses were in an intimate contact. Not much reaction from the cold-dynamic powers, even though they have an extra 1/27 cold power embedded in their masses intrinsically. It was an embryo universe at that cosmic time, but later when the universe grew bigger, things were changing rapidly. There was an interstellar space created where the cold powers were getting much stronger so that they were highly disputing the negatively charged kinetic masses in space which were moving around the central hot powers in the universe.

Having a higher potential now in power, they started to win in space by changing the elliptical orbit of the negatively charged objects of mass in the planetary systems to a spiral one. These are the same cold-dynamic powers pulling the galaxies towards the massive dark and crystallized mass of a ridge cold solidity of 100% in an absolute zero degree of temperature located in the outermost borders of the global universe. (See Figure No.22).

When the thermodynamic powers are prevailing, the weaker negatively charged kinetic masses in space will have no choice but to turn around towards their higher power sources until the cold-dynamic powers get stronger, and then a big war could start between these two heavenly almighty powers in order to be decided which one of the two will be the winner. From the beginning to the end, they are trying to find out which one is going to win those disputed negatively charged kinetic masses that are swinging around all over the space of the universe, like the 9 planets in our solar system turning around the sun. This is a general procedure that has to be followed during a complete cycle of our universe, lasting an expansion period of probably 2π billion years, and a contraction period of another 2π billion years. Also it is a law of the nature that has to be fulfilled by this intractable and evolutionary universe.

You may observe similar situations in the skies, watching the clouds and the satellites high over the earth's atmosphere being disputed by the thermodynamic power of the earth, and the cold-dynamic power of the interstellar space. The electrons spinning around their nuclei also are being disputed, being attracted at the same time by the nuclei of other atoms in their neighbourhood. I will recall that all these atomic planetary systems are only producing work with the electrons kinetic energy when the universe is still active and the average temperature in their surrounding space is above zero Celsius. See Chapter 18. Below zero Celsius the situation is reversed and cold-dynamic forces of attraction will absorb the hot energy that has been released before. Taking the whole universe as one planetary system as I have mentioned before, I can assume that all the kinetic masses are turning around the central core of an aggregation of stars and galaxies that you may find in the center of the sphere in our universe. At the same time, slowly and progressively these kinetic masses recede away from the center to the direction of the invisible wall of darkness and absolute zero temperature in the outer perimeters of the global universe (See Figure No.13c).

This recession is taking momentum when the volume in space has been increased enough, and the kinetic masses are approaching the invisible wall of darkness. This is because a higher (EMF) is created when the mean distance between thermodynamic and cold-dynamic powers are getting smaller and smaller, and the cold-dynamic force of attraction between them is getting higher and higher. The centrifugal forces in the whole universe are generated by the existence of the thermodynamic and the cold-dynamic powers in space. The kinetic masses are in the middle of the road and they are becoming an easy prey to be attracted and pulled around by any one of these hot and cold powers. But in different cosmic times, either the thermodynamic or the cold-dynamic powers have a different amount of strength. The whole action is illustrated in Figure No.22 by vectors, which represent strength of force and direction. These forces are of three kinds and it will remind us the mystery of the TRINITY.

These are: the force of attraction created by heat or cold powers looked upon as centripetal force, the second one is the centrifugal force generated by thermodynamic and cold-dynamic powers in space, and the third one is the resultant force of the negatively charged kinetic masses pulled around in space either by the positive or negative force of attraction developed by the thermodynamic or cold-dynamic powers. The negatively kinetic masses are the ones being disputed by both the great thermodynamic and the cold-dynamic powers. The extra cold-dynamic power of one part in every twenty-seven will in the END make this power to be the winner. The big war between the two super powers will start during the middle cosmic times of the universe.

Actually, there are two forces of attraction and two resultant forces of kinetic energy opposing each other as well as one centrifugal force, which belongs to both super powers. The centrifugal force is located in the middle and it is common for both these dynamic hot and cold super powers which are being diametrically located, opposing each other and working hard to see which one is going to prevail. They are fighting hard to see which one is going to win. The negatively charged masses located in the middle line between, are always the prey of the two super powers. Let me recall again that the hot and cold powers are condensed and pre-existed. The temperature in the open space either colder or warmer is created by necessity from the expanded or contracted heat power ratio of the 13/27 of the pre-existing mass. The negatively charged kinetic masses as one unit in the space of the universe are representing the same current (I) of the classic electrics in the equation of: ($I = E / R$). This negative current is a force of kinetic energy expressed as kilogram-meters of the whole package of mass in space revolving around its axis and at the same time circling inside and around the whole finite globular universe.

This total package of mass is the condensed mass of thermo/cold-dynamics in the whole universe. It is again a force of kinetic energy following a negative path since the beginning of the creation. The big bang was the commencement. It was the almighty power source of creation and energetic light emission in the square, which simultaneously created life, volume of space and kinetic forces. These forces were generated by necessity. They were the result of potential differences in space between the cold and the hot masses. And these potential differences developed **electro-motive forces and electric charges equivalent to coulombs** between the positively and negatively charged masses and finally these (EMFs) and electric charges created the forces of attraction which are the same strong, weak, gravitational and electro-magnetic forces **IN ONE**. In the space of the universe, since the commencement of the big bang there is a flow of negative kinetic mass, and we are all part of it. It is a negative path that we are all following. These total negatively charged kinetic masses are creating this current in order to fill up the vacuum which has been created by the big bang, sucking us all into the abyss of the metaphysical world before we all reappear again in another universe after another big bang occurs.

POSTSCRIPT

FINISHING THIS BOOK, "THE COSMIC THEORY OF THERMODYNAM-ICS", I should acknowledge that you may find a few shortcomings in the book while reading it as I interpreted to English from Greek, which is my mother tongue. I had no help from anyone while writing this book. It was not easy to unravel or follow out the sequence of the often arisen problems in this universe while I was writing this book. I wanted to follow the laws of the cosmic physics to the best of my knowledge in order to find the right answers. I believed that I was the only person that had these thoughts about our universe and our existence. I succeeded in my efforts to finish what I had in my mind for a long period of time of about 45 years to my complete satisfaction. After that, when I had this book ready, I made up my mind to go ahead and publish it. Problems are always present, but I always believed that this whole thing could end up successfully.

ABBREVIATIONS

G.U.T.	Grand Unified Theory
MeV	Million electro-volts
(∝)	Infinity
(tM)	Total Mass
EMF	Electro-motive Force
(R) Meaning in Electricity	Resistance in Electricity
(R) Mathematical Meaning	Radius in the Circle
(I)	Current in Amperes
(mDen)	Mean Density
(mT)	Mean Temperature
(e⁻)	Electron
(e⁺)	Positron
(P)	Proton
(Vol)	Volume
(Vol-1)	Volume 1
(Vol-2)	Volume 2
(v)	Neutrino
*	Multiplication
(N)	Neutron
(He)	Helium
(Ac)	Actinium
(HP)	Positive Heat Power
(hp)	Negative Heat Power
(CP)	Positive Cold Power
(cp)	Negative Cold Power
(Dist)	Distance

(t)	Time
(A) Mathematical Meaning	Abstract Number and Quantity of Anything
(E)	Potential Difference or Voltage in Electricity
(Ke)	Kinetic Energy
(c) Mathematical Meaning	Speed of Light
(el-ch)	Electric Charge
(FATT)	Force of Attraction
(mDist)	Mean Distance
(mRd)	Mean Radius
(n)	Nucleus
(Rst)	Resistance
(tW)	Total Power